LIFECODE

L I F E C O D E

The Theory of Biological Self Organization

Stuart Pivar

Ryland Press

LIFECODE

TheTheory of Biological Self Organization

Ryland Press

Library of Congress Cataloging-in-Publication Data

LIFECODE

TheTheory of Biological Self Organization

Stuart Pivar

Ryland Press
New York, NY

ISBN 0-9749860-0-3

Library of Congress Control Number: 2004101500

PLATO

"The Book of Nature is written in the characters of Geometry."

GOETHE

(in a letter to his sister)...

"Tell Herder that I am extremely close to the secret of the generation and organization of plants and that it is the simplest thing that one can imagine... the same law will lend itself to all other living things."

Contents

Plates

Foreword

The recent passing of Francis Crick was noted by the recounting of the discovery of the structure of DNA and the incalculable benefits accruing from the new science of genetic engineering. But there was no talk of the secret of life, or the blueprint for the body, which for decades after 1953 was the slogan of genetics, when thousands of biologists addressed themselves to the cracking of the DNA code for the construction of the body. All efforts since then to associate gene sequences with the production of actual form have failed. The genome is now widely believed to function as systems of switches, which trigger other mechanisms, not yet identified, which shape the body. The genes provide the traits of the parents as to the relative size of the parts of the body, the chemistry of the immune system, and superficial characteristics such as skin, eye and hair color. The solution to the riddle of morphogenesis, the system by which one cell becomes the complex body, has remained a secret.

The secrets of force and motion were revealed by Newton. Mendeleev's Periodic Table of the Elements led to a rational theory of chemistry. Biology has no rational theory. There is no understanding of development and evolution, the science of how the body is made and evolves. Few biologists still believe that Darwinism plays a fundamental role in the creation of body form. Yet work continues in a relentless effort to decipher a code for the body by comparing the genome of fruit flies, worms, mice and people. The idea that form is the result of a cascade of gene effects of unimaginable complexity is now the basis of the genetic theory of form.

Many biologists have long suggested that the origin of organic form lies in the self-assembly of unseen patterns or structures in the egg cell. But no model system with actual form-predictive power has been proposed heretofore.

This book presents one. It makes the hypothesis that:

The animal and plant body is formed by the direct enlargement of the membrane that encloses the egg cell and is shaped by an ingrained, self-organized pattern.

The membrane that forms on the surface of the microscopic speck of primordial germ plasm that begins the formation of the egg provides the substance for the body it will produce by expansion and later by subdivision into individual cells.

The self-organized pattern that occurs in the surface serves as a template, shaping the growth of the complex mature egg and, after subdivision into cells, continuing to direct the formation of the adult body form.

The shape assumed by the surface enveloping the primordial germ plasm is a geometric or topological form called the sickle torus, a kind of torus that can be described as one spheroidal surface inside another, interconnected at the poles by a narrow funnel vortex. Its cross section is sickle-shaped as opposed to the circular cross section of simpler doughnut-shaped torus.

The toroidal egg cell surface is fluid. It is in constant axial flow like smoke rings or magnetic fields but substantially slower.

Toroidal streaming imposes on the membrane sheet the relentless cycle of alternate compression and expansion suffered as the sheet passes through the narrow vortices. The membrane becomes empatterned by radial and circumferential subdivisions. This pattern guides the expansion of the membrane to the mature egg-cell plasma membrane and later, as separate cells, guides the formation of the standard adult phyletic body.

As the egg grows in volume, the toroidal surface expands. With expansion it must undergo one of several possible topological deformations, depending on its fluid properties. These account for the phyletic niches.

Given the standard body plan for the phylum, the genetic inheritance code carried by the genome distinguishes clam from oyster, man from mouse, man from woman and individuals from each other.

Each body is constructed by the expansion of the self-assembled surface of the membrane of the egg from which it arose (Plate 1). All plant and animal body forms derive from this structure.

The theory presented here implies a new science of biology, based on the discovery that a self-organized, geometric structure can generate the universe of phyletic body forms, and that development and evolution may be accounted for graphically by the predictable topological events proceeding from its growth.

The ancient Riddle of Morphogenesis asks how a single, nearly invisible egg cell can generate the complex adult body. Johann Wolfgang von Goethe, who coined the term "morphology," asked from what single, simple regular structure can all the plant and animal forms be derived, which are so universally recognizable as living. Like snowflakes, so different and yet so alike, they are related, as are all crystals, only by common morphological origin.

Homunculus is the name of the miniature man that the medieval and renaissance alchemists attempted to concoct from elemental substances (Plate 2). It is also the name later given to the miniature man that the seventeenth-century microscopists thought they saw in the egg. They were not able to propose plausible theory as to how it got there. This new theory may be seen as a homunculus theory, where the homunculus is not a mini-micro-man, but a molecular structure that forms spontaneously in the egg.

As a mathematical or geometric theorem, the omni-generative power of the streaming torus is presented as proof by demonstration of congruency, that the sickle torus surface can produce phyletic body shapes by simple deformations. Whether this hypothetical model is indeed the mechanism in use by nature is another question. If this can be reasonably demonstrated, then the accepted theory of biology that present forms resulted from the natural selection of random mutations in a genetic code that directs the formation of the body is all wrong.

With this new view of biology, the forms of plants and animals are not randomly chosen from chaos like a pile of stones, rather living nature is a perfect platonic universe where each species is related to all others through a common geometric ancestor. Life elsewhere must be like ours. The large-brained vertebrate is a biological inevitability.

The amazing phenomenon of the virtual identicalness of all individuals of the human race in body, behavior and thought can be comprehended as a phenomenon of literal material transubstantiation, in which each cell of the individual comprises substance from the single egg cell derived from the mother, the mother's mother, back to Eve. The communality of brain cells provides the infrastructure for communal consciousness.

PLATE 1 *(page 12–13): The Toroidal Surface Expands. With expansion it must undergo one of several possible topological deformations, depending on its fluid properties. These account for the phyletic niches. All plant and animal body forms derive from the torus. Each body is constructed by the expansion of the self-assembled toroidal surface of the membrane of the egg from which it sprang.*

PLATE 2: *The miniature man the medieval and renaissance alchemists attempted to concoct from elemental substances was called the Homunculus.*

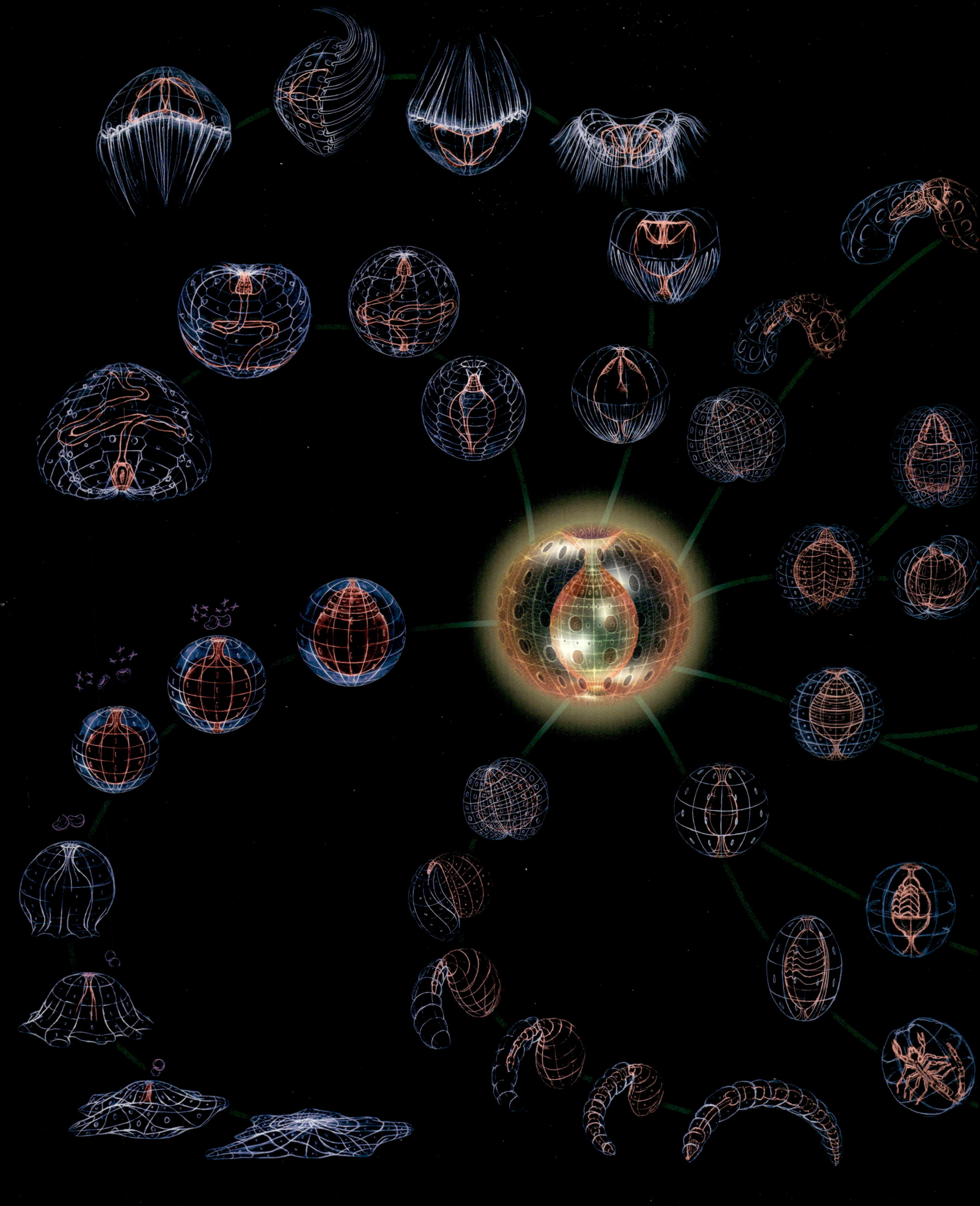

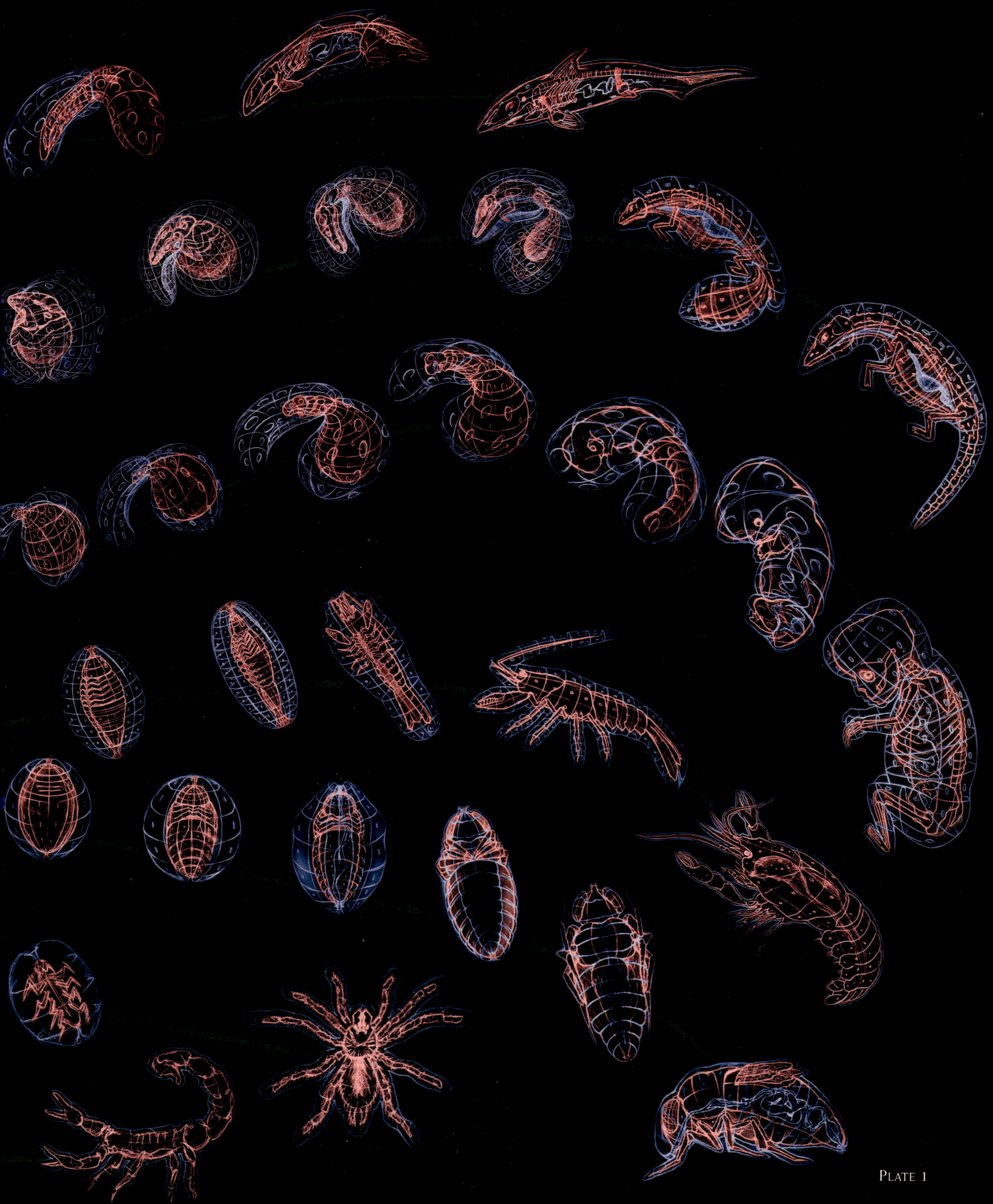

PLATE 1

Chapter 1 The Origins of Species by Other Means

An amoeboid wanderer cell grows as it goes, finally rounding up as a swollen ball. The expanding surface causes opposite poles to indent, meet in the middle, forming a canal. The surface commences to flow into the north pole and out the south, relentlessly. The consequent cyclical compression and expansion of the surface generates an ingrained pattern of folds and segments. The surface folds on entry into the vortex, and on exiting, expands, fanning out over the interior surface.

The surface flowing into the vortex is subdivided by lateral compression into several component, coaxial vortices. These may number two, three, four or five. These may fuse axially, forming tubes that may remain sealed or may open, spreading laterally on the spherical interior surface, forming wave patterns like the wake of a boat, the whorled pattern of a rose, or the undulating tentacles of a medusoid jellyfish.

Or the surface may split from pole to pole and roll inward along the seam. The surface may segment, buckle and fold forming jointed appendages as of insects and crustaceans.

Or lastly, a split may result in the ingression of one hemisphere into the other, dragging the interior surfaces to the outside, patterned as the hide of a vertebrate.

The stretched fabric finally tears, instigating an elastic recoil and sudden deformation of the surface. The dynamic contour is preserved as the fluid form suddenly congeals in one of a small number of topological possibilities, like a glass bottle at the end of the glass blower's pipe.

This abbreviated statement and the illustrations and charts, elucidated by the narration, are designed to offer intuitive comprehension of the simple premise of a new theory of biology.

This theory may be viewed as a solution to a mathematical problem, namely how to generate the complex biological body form from a single cell. The solution is presented as a demonstration of congruency of the forms predicted with those of observed nature. But predictive success is not proof that the model system is the one actually in use by nature. Since the progenitor molecule itself has not been directly observed it remains a hypothetical construct designed to produce a desired result, much like the DNA molecule or the atom.

Form without Inheritance

This new account of morphogenesis is based on the hypothesis that the egg cell is not a sphere, as is generally supposed, but is a torus, consisting of the streaming surface

PLATE 3: *The Torus*

Plate 3

of a sphere with internally interconnected poles. Smoke rings, mushroom clouds and magnetic fields are other examples of the streaming toroidal surface. Topology distinguishes between two fundamental enclosed curved surfaces, the sphere and the torus. The toroidal surface behaves very differently than the spherical surface. A spherical surface is uniformly curved. An indenting force is met with resistance. However, an ingression in a streaming toroidal surface may initiate an internal flow, forming a new internal structure. The spherical surface is in stable equilibrium possessing inertia of rest. A point displaced by a force will restore its position. The spherical surface cannot form a continuous stream. The toroidal surface is in unstable equilibrium possessing inertia of motion. Once started there is little resistance to toroidal streaming.

The hypothesis that the cell is a torus, the principle contention of this theory, is consistent with the established understanding that the endoplasmic reticulum, the major organ of the cytoplasm, is an extension of the cell membrane and the nuclear membrane, and therefore may well serve to interconnect the poles. The following evidence supports the theory that the cell is a torus.

1. The direct observation in early oögenesis of characteristic internal, toroidal streaming.
2. The cell membrane behaves like a toroidal surface seen in the automatic formation of vesicles and cyto-skeletal structures.
3. The membrane is an expanding bilayer. Such surfaces have the property of automatically forming a streaming toroidal surface and of assuming the contour of a torus.
4. It has been demonstrated that artificial cells generated from phospholipids assume the form of a torus.
5. The body plan of plants and animals is based on a streaming torus.

The egg cell membrane can be imagined as a diaphanous, labile bilayer torus bubble capable of maintaining its patterned surface under extreme deformation. In its near-zero gravity, near-zero friction, molecular-size world the cell membrane flows like water. The characteristic contours of living matter derive from this early liquid phase of cell formation. That bilateral animals are universally hydrodynamically shaped for anterior locomotion is mere coincidence. Blown glass bottles are similarly shaped by the congealing of a fluid form.

Oögenesis, the development of the animal egg, or Oöcyte, begins as an amoeboid cell in the female embryo, which travels the body toward the gonads and eventually becoming a newly formed spherical egg. You begin inside your grandmother's uterus.

At this stage, the egg is seen by this new hypothesis as a streaming, radially symmetrical, continuous surface in the form of a globe within a globe, interconnected at the poles by a small funnel vortex (Plate 3). The surface streams in at the north pole, spreads out internally, flows downward, exits at the south pole and ascends externally to repeat the cycle. The cycle of compression and expansion experienced by the toroidal surface generates a pattern consisting of a rectangular grid, which undergoes continuous cyclical transformation creating a radial-axial pattern (Plate 4).

PLATE 4: *Toroidal Streaming*

PLATE 4

The egg surface is continuously expanding. Eventually the highly stretched fabric gives way at a weak point, or along a seam. The surface flows in rapid rearrangement in response to the catastrophic event of puncture.

To define form means to establish its geometry by measuring the distance between points on its contour. But biological form is a dynamic, living system, constantly changing like a fountain or the blowing of molten glass. The streaming toroidal surface is best defined by topology, when the character of the labile, elastic surface is considered, regardless of its momentary geometric configuration. The biological body has the dual characters of the topological and geometrical. The dualistic character of living tissue as either individual cells or a whole membrane derives from the phenomenon of a geometric-topological duality.

The character of the streaming toroidal bilayer may be derived from the eternal instability of its surface, one side compressed, the other side stretched, each seeking the state of the other in a continuous effort to turn itself inside out. This behavior is exemplified in each of the small variety of topological solutions reached by slightly differing egg cell patterns in the urgent quest for dynamic stability following the catastrophic rent in the fabric. Each solution is a phyletic niche.

In radial animals the membrane separates as it tears evenly about the south polar axis. The cell then takes forms like an umbrella, from jellyfish to starfish, by turning inside out at a point.

The bilateral animals turn inside out along a line. They are the result of a meridianal split in the fabric leading to a meridianal incursion. In the arthropods this causes the compressed segments to buckle and fold inward, molding themselves into the pairs of hollow, jointed legs of insects and crustaceans.

The vertebrate body results as one hemisphere of the meridianally cleaving egg surface enters into the interior of the other, creating the body architecture in one pass, the bones forming as the transparent segments are transformed by the passage through the inversion vortex, the whole shrink-wrapped in skin, during the embryological event known as gastrulation. These forms become visible only later, upon ossification.

In plants, the response to the tear in the egg cell membrane is the ingression of the south pole by an internally directed conical shoot that penetrates the cell and emerges clear out the opposite pole, drawing the reticulate pattern of its outer surface into cylindrical form, later to reappear as flowering branches which repeat the toroidal pattern.

The Irrelevance of Cleavage

These descriptions of embryological origins are based on the properties of a single cell-encompassing membrane. But the pre-embryo blastula consists of a thousand cells. Embryology traditionally consists of tracking the path of these cells, forming the embryo as they proliferate. A century of this analysis has proved fruitless, except to provide detailed descriptive data for a process without the slightest understanding as to what is going on. The error lies in the rejection by biology seventy-five years ago of the organicist theory, which considers the organism as a whole, incidentally subdivided into cells. This organicist view leads to a single solution to the topological problem of morphogenesis, the streaming torus.

At fertilization, the egg proceeds to subdivide itself in a series of cleavages resulting in the blastula, the hollow ball of cells from which the embryo is formed. A premise of the automatic biological construction theory is that the surface of the blastula is a mosaic of the membrane of the egg cell, and that the blastula surface in forming the embryo acts as a single membrane, preserving the configuration of the cell membrane form subdivided.

The first cleavage leaves two cells, each possessing half the original egg cell membrane. Next there are four, each with a quarter of the original membrane. With many cleavages a mosaic of the egg is produced, each cell bearing a fraction of the original membrane. Thus, gastrulation and organogenesis are directed by the deformation of the empatterned membrane of the cell rather then by the agendas of the individual cells of which it is actually composed. The embryo begins as a subdivided membrane. Only much later it becomes a community governed by functionally different cells.

This theoretical account of the formation of the vertebrate embryo describes the stages of embryogenesis that follow the formation of the thousand-cell blastula resulting from the subdivision of the egg. Based on the nature of a sheet of early cells to act as a membrane, events are described as if they are undergone by the egg membrane rather than the tissue of cells of the blastula. This hypothesis of the dual character of the expanding surface of the blastula is fundamental to the theory of self-organization. The failure to consider this phenomenon is responsible for the obscuring of the workings of morphogenesis until now.

Cleavage results in subdivision, which is the basis for the system that results in the enlargement to immense proportions of the embryonic form by the mosaic proliferation of the individual cells. Otherwise, cleavage is irrelevant morphologically and serves to obscure the understanding of the process, especially when cleavage occurs simultaneously with gastrulation.

In the 1950 publication of the *Natural History of Plant Form*, Agnas Arber addresses the problem of the dual source of botanical form in the seed and in the bud. She makes a clear statement of a "holistic conception of plant morphology," declaring that the forms generated at the tip of each shoot are directed by "an original pattern in the relevant part of the embryo." (See Chapter 19)

Chapter 2 **Vertebrate Morphogenesis**

PLATE 5: *Vertebrate Body. The dorsal exosphere generates the vertebrae and ribs; the ventral exosphere, the chest; the ventral endosphere, the body cavity and blood circulation system. The whole is enveloped in the skin; the former interior surface pulled inside out during the process.*

Just as there are examples in nature of crystals in each of the geometrically possible configurations of faces and angles, there are phyletic body forms in many of the imaginable ways a toroidal balloon can be expected to react to a puncture. The most complicated of these is the vertebrate. The vertebrate body is formed by an assembly process of a few hours as sections made at diverse locations on the membrane during oögenesis come together during gastrulation.

The vertebrate Homunculus consists of a toroidal polyhedron subdivided by twenty segments on an external and internal spheroid, the whole subdivided radially in eight sectors producing a toroidal polyhedron of 320 quadrilateral faces. Each by topological transformation will generate a part of the body. Later, this established dynamic system of morphological transformation provides the structural and hydrodynamic basis for the blood circulation system of the adult (Plate 8).

The southern vortex tears, instigating a sudden elastic release as the entire figure recoils upward. Three segments compress, forming a separately delineated cap. The segment bands below it bulge from the lateral pressure. The direction of toroidal flow is arrested and reversed, as the axis advances anteriorly, that is upward, while the outer surface retreats downward.

The vertebrate egg splits along a meridian, dividing each sphere into two hemispheres, the dorsal and the ventral, named for the contribution each will make to the assembled body. During the division of the cell, the nucleus becomes enclosed in the ventral hemisphere. The dorsal endo-hemisphere folds meridianally inward entrapping a tube that will become the alimentary canal. A dome is formed as the three uppermost segments of the endosphere are compressed by lateral pressure of the elastic response to puncture. The dorsal and ventral center sections of the dome fold inward and become sealed, as the future brain case and palate of the future head.

The process of inversion, gastrulation, begins with the entry of the head segments into the interior of the dorsal hemisphere, pulling the ventral hemispheres along as well as the alimentary tube enclosed in the dorsal endo-hemisphere. The dorsal exosphere generates the vertebrae and ribs, the ventral exosphere generates the chest, and the ventral endosphere generates the body cavity and blood circulation system. The whole is enveloped in the skin, the former interior surface pulled inside out during the process (Plates 5, 6, 7, 8).

PLATE 5

Shaping the Bones

All leading edges of the involuting tissue continue to roll as they enter the mouth of the vortex at the outset of gastrulation, concurrently enwrapped by the sheets of membrane, which will later become muscle (Plate 7). The ends of the rolled cylinders of tissue are deformed by forces of compression, bending and twisting, which will later produce the ball and socket, and knee joints.

Shape is imparted to the members, or organs, in the course of the dynamic process of flowing tissue. The contours of the gastrulation vortex create the form, as they are frozen in mid-motion.

Adult form is defined at the moment the flowing figure arrests plastic deformation, commencing growth without further development (Plates 5, 6). That moment in development must be controlled by the genome, which can cause its advance or delay.

Segments are shaped by axial rolling produced by the forces of compression and tension in the reversal of curvature of their inner and outer surfaces. The undersurface of the segments flares out at the ends, generating either a convex or concave curl, forming a ball or a socket, remaining cartilaginous, and forming crescent-shaped pads or flat discs that engage end to end between the bones (Plates 6, 12, 13, 14). All bones are variants of this fundamental topological maneuver. Fabrication of the musculo-skeletal system begins as the shaped bones are overlaid with the cortex surface. The protein strands attach to the protuberant bone surfaces forming muscle tissue.

Attachment of muscle to bone occurs after the segments have folded under compression, forming cubical stacks, the faces of which are the exposed ends of the bones. The faces become attached to the adherent sheet of protein strand. Upon unfolding, the presumptive muscles form bundles that envelop the bone and are attached at or near the ends (Plates 7, 15).

The Skull

The head forms from the polar cap, which consists of three flat, broad segments of the polar sphere appearing as three concentric, ringed zones (Plates 9, 10, 11). The segments are divided into four quadrants. Each of the resulting twelve parts develops a circular hole at the center. The six fenestrae of the future dorsal side will form the nasal, optic and auditory capsules. The ventral half produces the jaw. The architectural scheme of the skull results as concurrent, deep, invaginating infolds along the dorsal and ventral midlines fuse, sealing off enclosing cavities in each hemisphere. These meet internally and form the floor of the brain case and the roof of the mouth. These cavities are analogous to the structure in plants that produce the anthers.

From this universal homuncular figure the precursor of the adult skull is formed by a single maneuver, the reversal of direction of the toroidal flow of oögenesis. The axial centerline of the dome is forced to stream anteriorly. The exterior surface shrinks, causing a series of simultaneous effects that outline the adult skull: the segments bulge, the rear teeth compress, the jaw hinge forms by the twisting of the parietal segment, the dorsal surface peels back forming the zygomatic arch, the basi-occipital compresses around the spinal cord.

PLATE 6: *The Vertebrate Skeleton*

The brain is an enclosed membrane within the brain case. With expansion, it

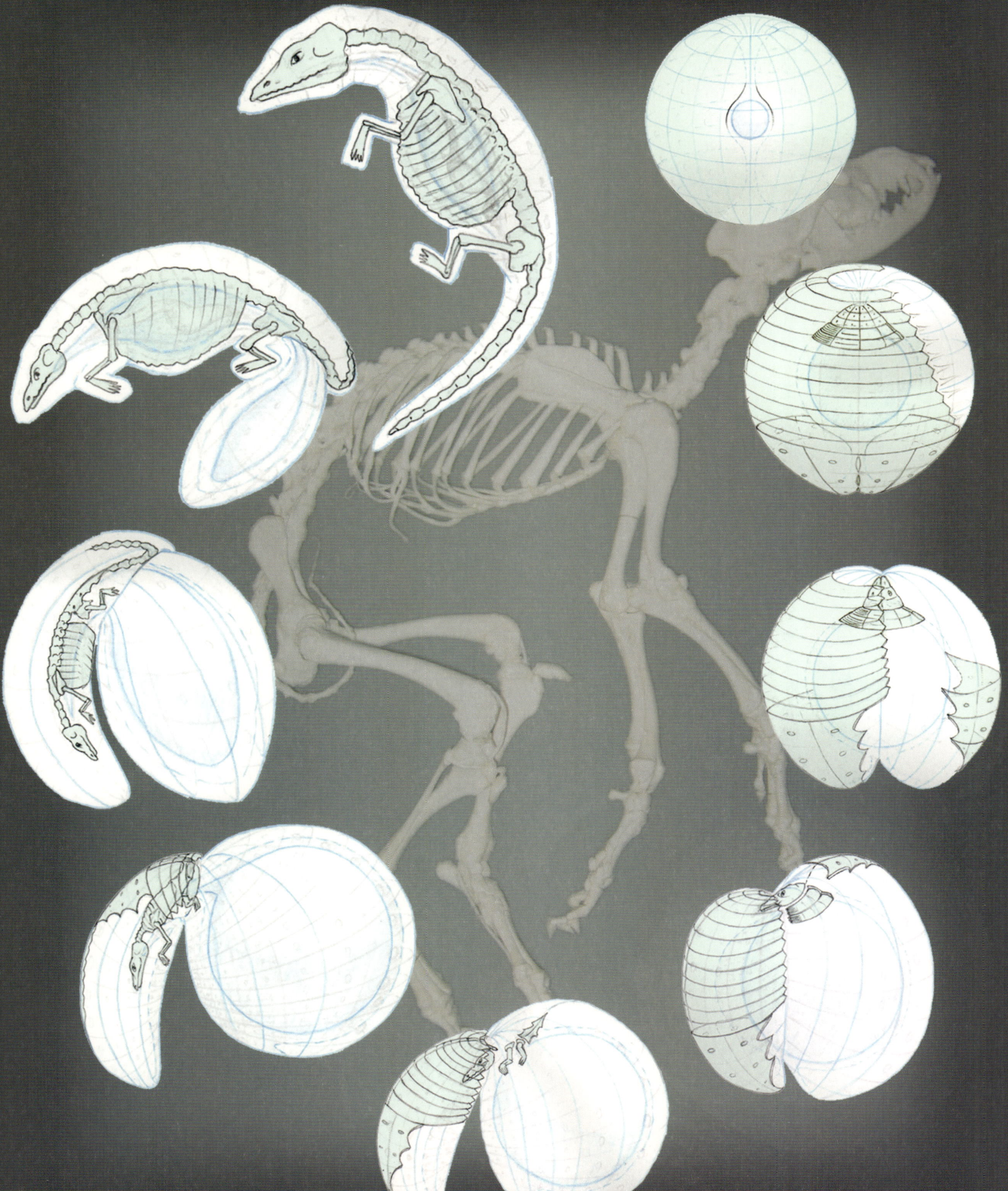

PLATE 7: *The Origin of Vertebrate Muscle. All leading edges of the involuting tissue continue to roll as they enter the mouth of the vortex at the outset of gastrulation, concurrently enwrapped by the sheets of membrane, which will later become muscle. Attachment of muscle to bone occurs after the segments have folded under compression, forming cubical stacks, the faces of which are the exposed ends of the bones. The faces become attached to the adherent sheet of protein strand. Upon unfolding, the presumptive muscles form bundles that envelop the bone and are attached at, or near the ends.*

PLATE 8: *Fluid Circulation in the Vertebrate*

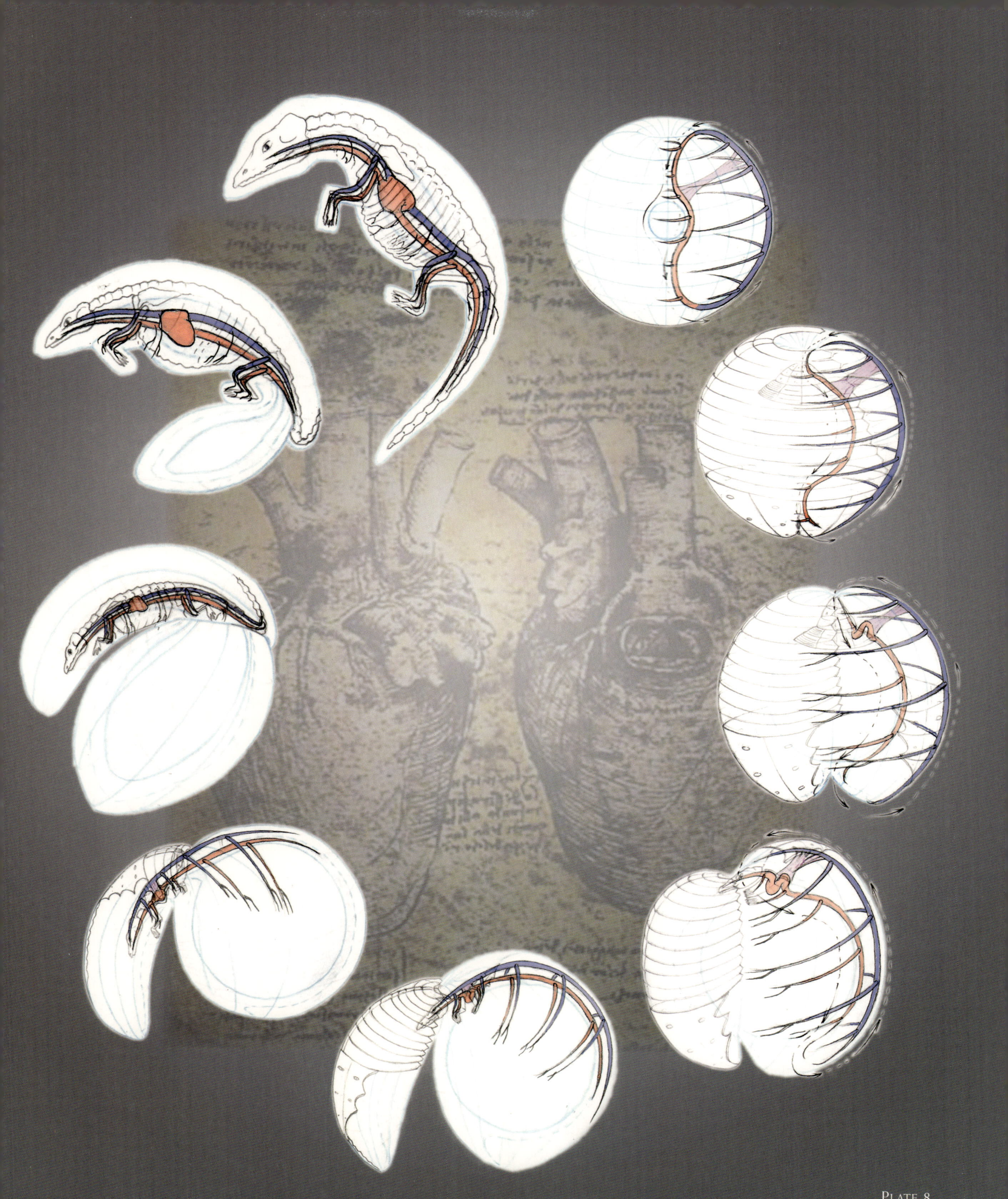

PLATE 9: *The Formation of the Skull*

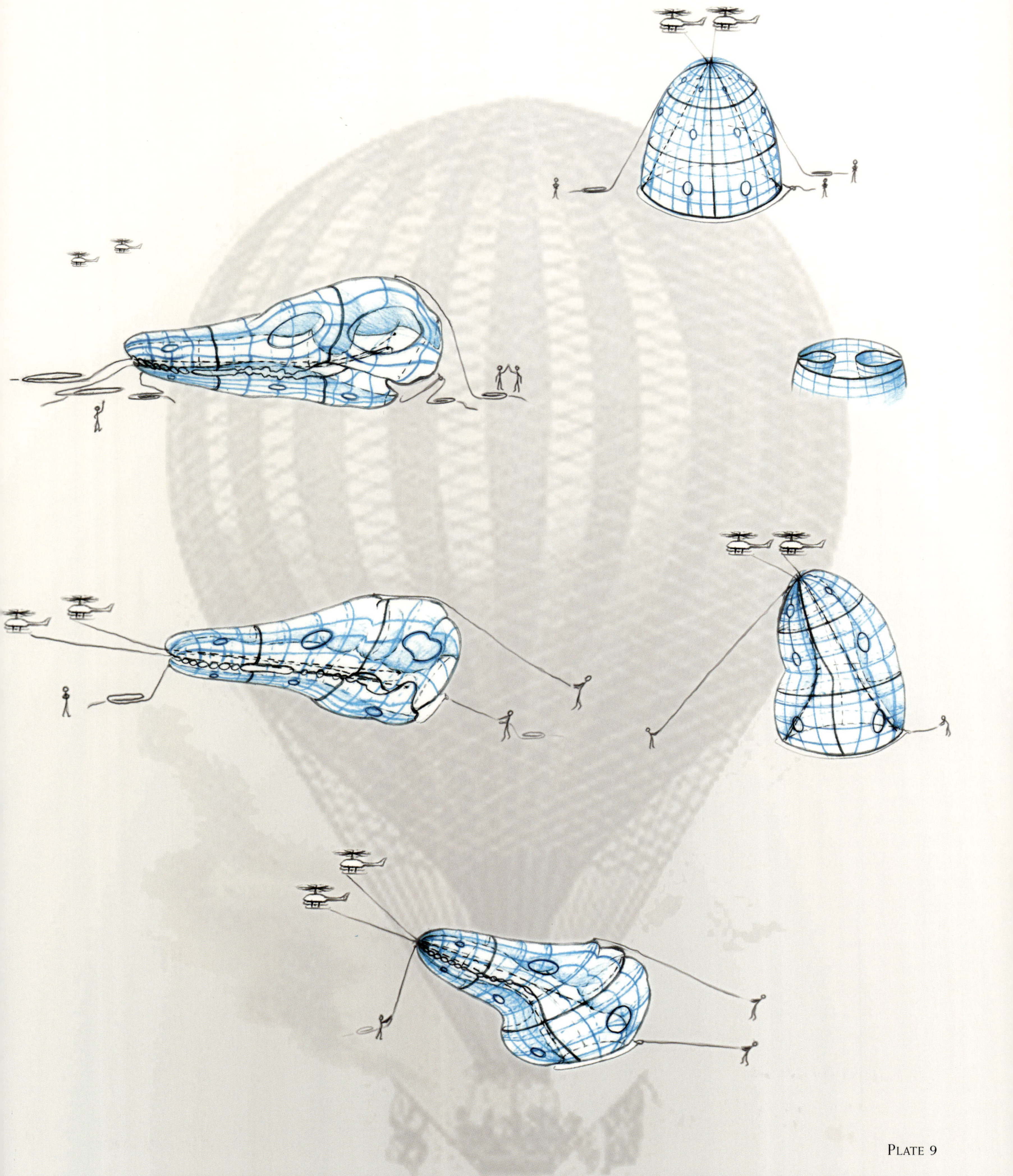

Plate 9

PLATE 10: *Cheetah Skull and Brain. The brain is an enclosed membrane within the brain case. With expansion, it bulges and extrudes through six fenestrae, extruding to form nasal, optic and auditory sacs.*

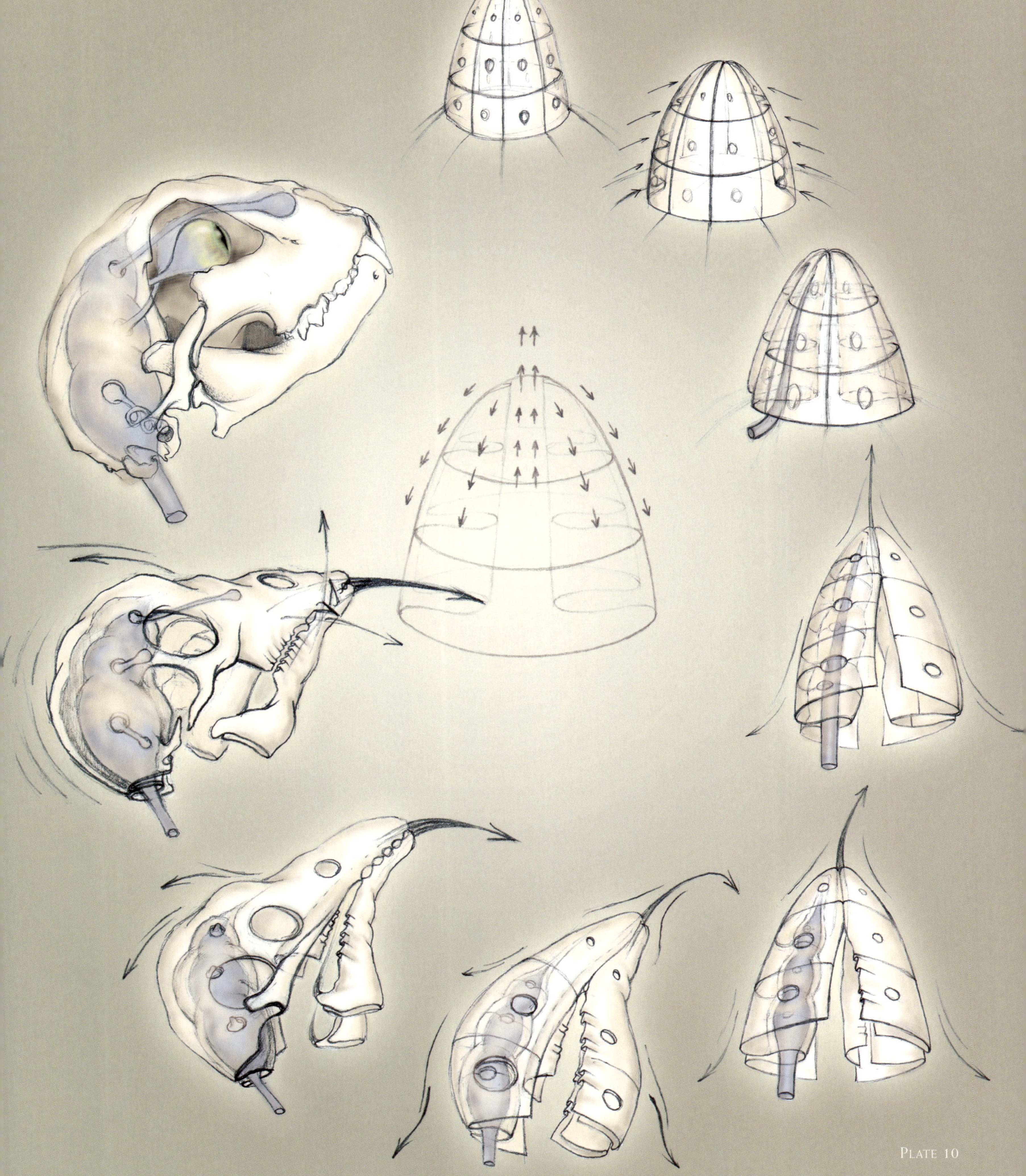

bulges and extrudes through six fenestrae, extruding to form nasal, optic and auditory sacs (Plates 9, 10).

The careful concordance of this brief description of the topological event of skull formation with the diagrams will enable the rationalization of every structure or feature of the skull.

Vertebrae

Material at the midline, headed toward the inversion port, is pressed together forming a tube as it enters the aperture, converting bar-shaped segments into a row of spools, separated by discs (Plate 39).

Rib Cage

Distal segments reverse curvature as they pass through the outlying part of the vortex, folding into bales, which line up in pairs alongside the forming vertebral column. These unwrap and elongate, forming the tissue for the future ribcage (Plates 5, 6).

The Sacrum and Xiphoid

The southern polar cap splits to produce the sacrum dorsally and the xiphoid process ventrally. The sacrum structure is subject to extreme distortion in its formation (Plate 12). It is unrecognizable or absent in many species. The primate sacrum retains the archetypal form.

The origin of dorsal fins of fishes, the subject of conjecture by nineteenth-century morphologists, may be seen as the result of the dorsal compression of the sacrum, the posterior homologue of the skull.

Limb Girdles

Compressed segments of the limb girdles stretch and part at the ventral midline, forming limbs. This separates scapula, humerus, radius and ulna, and the carpals. Telescopic separation forms the limbs as the laminae of the surface of the coaxial cylindrical surface separate serially under tensile failure (Plates 13, 14). The complete or partial separation of radius and ulna results from the axial split of the segment during compression failure. Similarly the axial folding of the carpals forms the opposable thumb.

Fingers

Fingertips, originally attached to the future cervical vertebrae, now move far from the point of attachment, remaining connected by elongating cables of membrane. Axial segmentation of the digits results as the surface laminae split consecutively in a telescopic failure pattern. They remain attached by a long tendonous membrane, which comes to lie along the length of the limb, maintaining a direct connection from fingertips to the neck, the basis of the musculo-skeletal system of the limb (Plate 16). The same phenomenon of telescoping, coaxial laminae accounts for the row of diminishing graduated segments of the tail.

Axial tension during inversion also shapes the pelvis and the scapula, resulting in stretching and the tearing of the membrane through the circular holes at the center

PLATE 11: *Skull Formation*

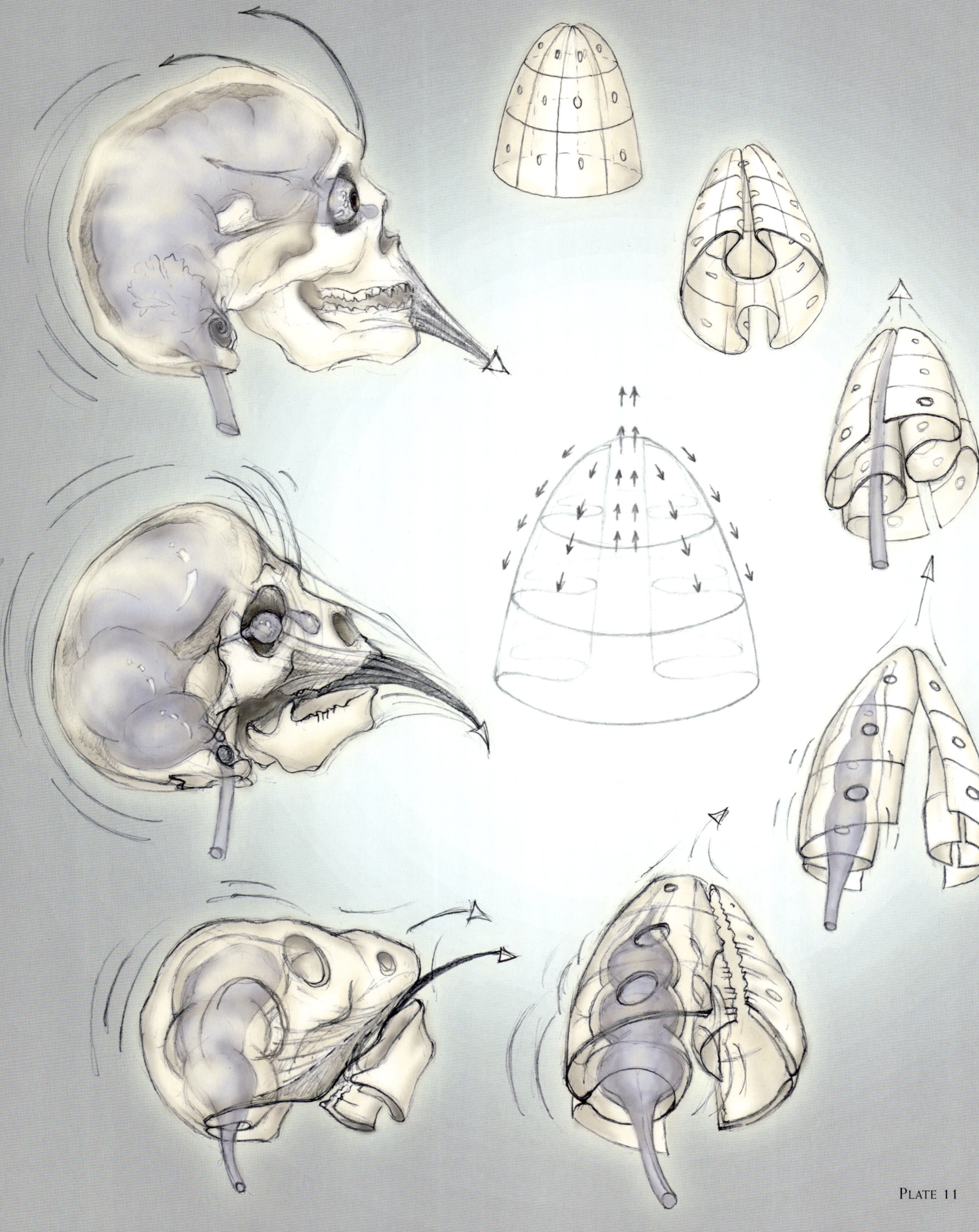

PLATE 12: *Sacrum and Xiphoid. The southern polar cap splits to produce the sacrum dorsally and the xiphoid process ventrally.*

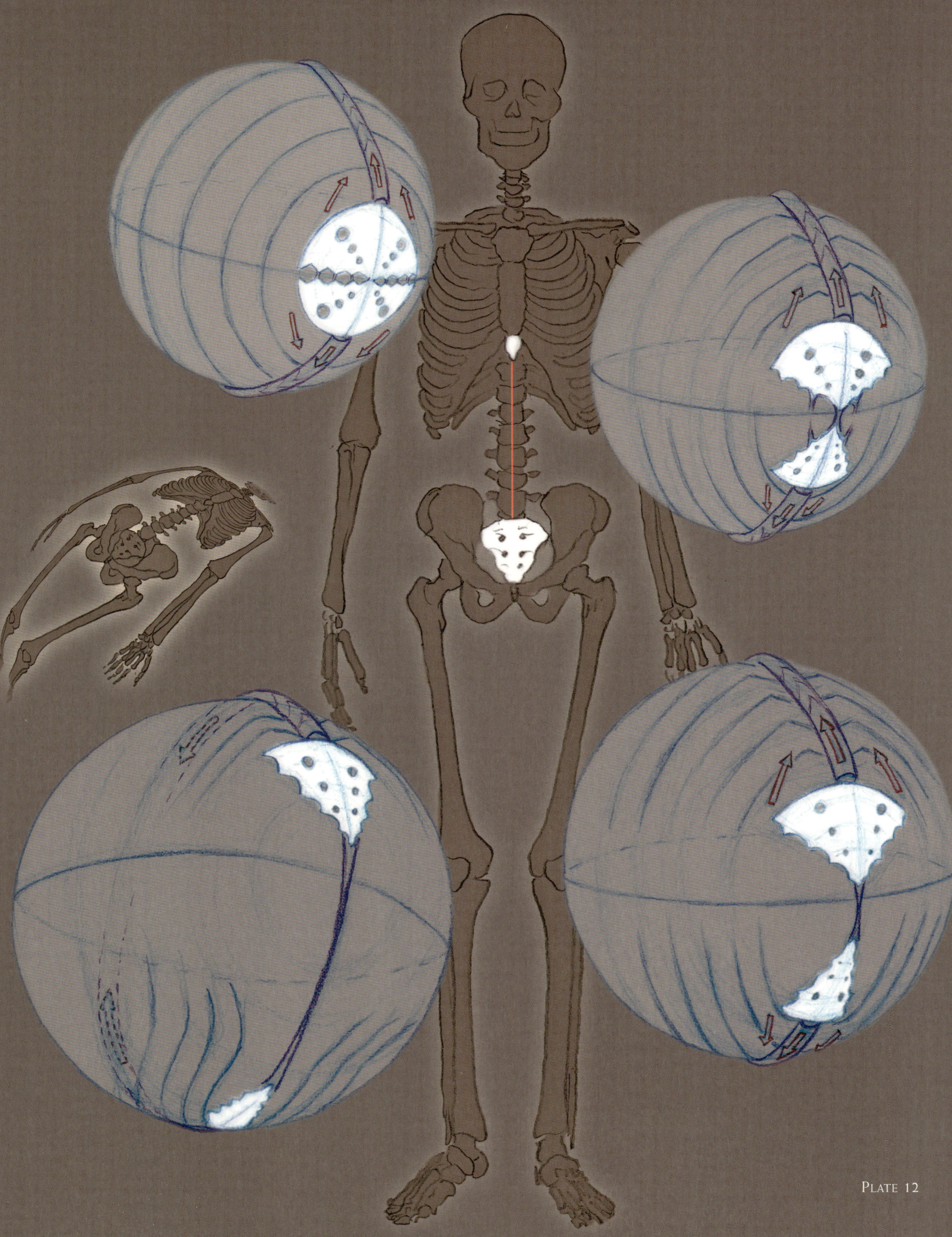

PLATE 13: *Limb Formation*

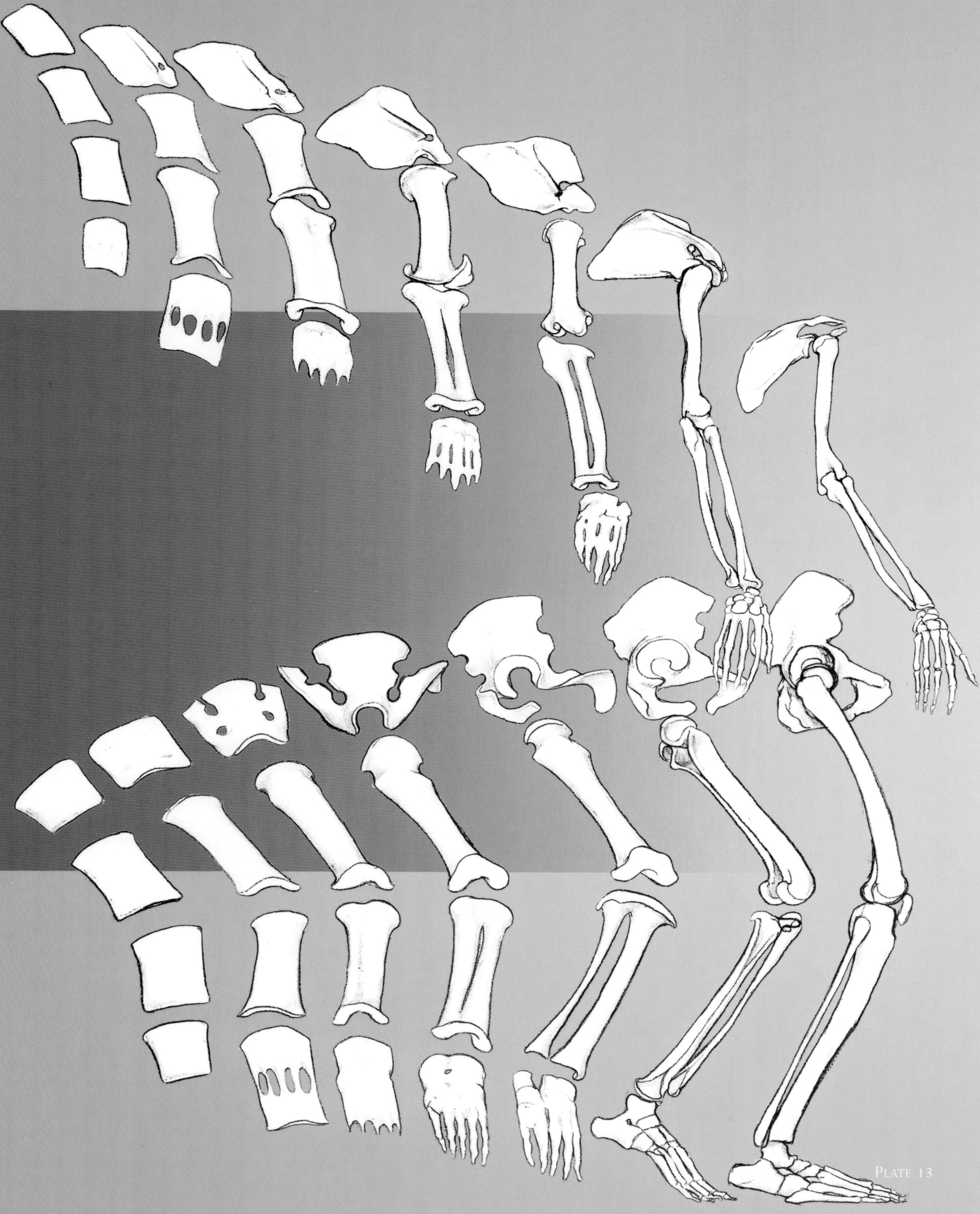

PLATE 14: *Pelvis and Scapula. Axial tension during inversion shapes the pelvis and the scapula, resulting in stretching and the tearing of the membrane through the circular holes at the center of the segments. Extreme stretching of the two sides of the pelvis creates the ischium and pubis.*

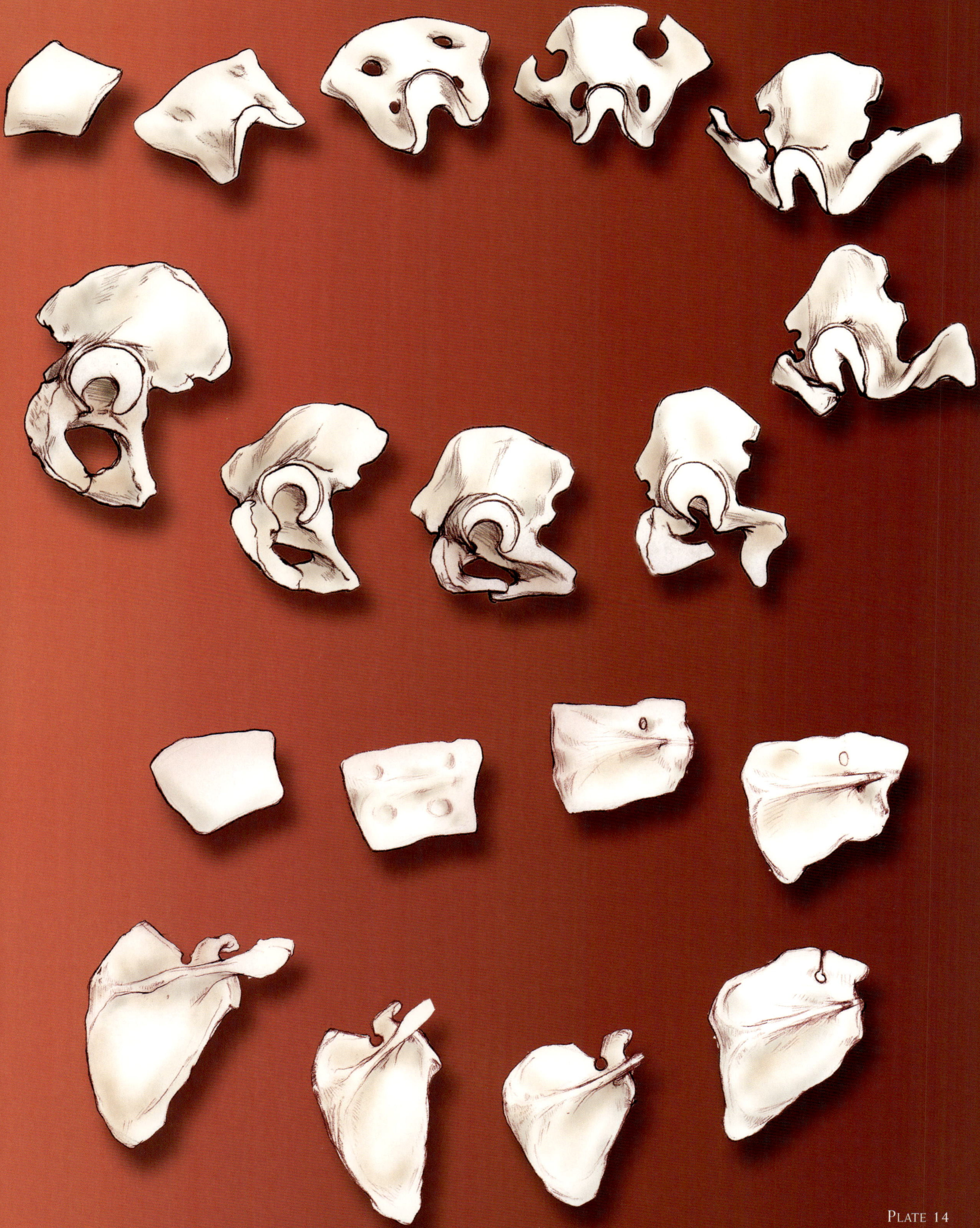

PLATE 15: *The Musculo-Skeletal System. Attachment of muscle to bone occurs after the segments have folded under compression, forming cubical stacks, the faces of which are the exposed ends of the bones. The faces become attached to the adherent sheet of protein strand. Upon unfolding, the presumptive muscles form bundles that envelop the bone and are attached at or near the ends.*

of the segments. Extreme stretching of the two sides of the pelvis creates the ischium and pubis (Plate 14). The ischium may remain in the anterior, or be distended along the posterior. This is the well-known simple difference that distinguishes saurischian from ornithiscian dinosaurs. The substance of the scapula is drawn out to form the acromium and the clavicle. In birds, this proceeds farther, resulting in a large sternum with only a thin blade of a scapula remaining.

The musculo-skeletal system originates as contiguous points separate, remaining connected by a thin, tubular membranous thread. The attachment and insertion points of any muscle were thus once together, separating in elastic strands. The attachment of the brachial nerves to the cervical vertebrae confirms that the fingertips were originally each attached to one of them, and similarly the toes to the lumbar vertebrae. The fingertips appear to originally attach to the sternum vertebrae, or sternabrae, then migrate dorsally and fuse with the cervical vertebrae. This distinguishes the morphology of the cervical vertebrae from the other types. It is especially noticeable in the juvenile human where fusion of the two parts of each of the cervicals is incomplete (Plate 17).

The vertebrate tail is morphogenically distinct from the vertebral column and the sacrum, although these three are seen to fuse in many species. Tail segments result from the phenomenon of telescopic segmentation as described above for fingers.

The origin of eyes and sex organs can be accounted for by the fate of the centrioles, which disintegrate after mitosis. Eyes and the genital organs have anterior-posterior morphologically symmetrical homology and similarity in construction, each comprising spherical vesicles, the interior surface lined with active bodies, either retinal cells or germ cells. The vesicles migrate distally to the exterior in eyes and testicles.

PLATE 17: *Digit Formation. Axial segmentation of the digits results as the surface laminae split consecutively in a telescopic failure pattern. The same phenomenon of telescoping, coaxial laminae, accounts for the row of diminishing graduated segments of the tail.*

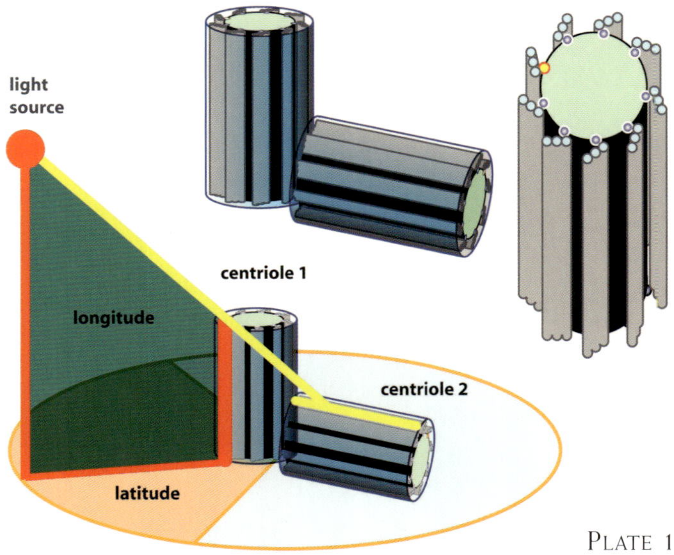

PLATE 16
Centrioles

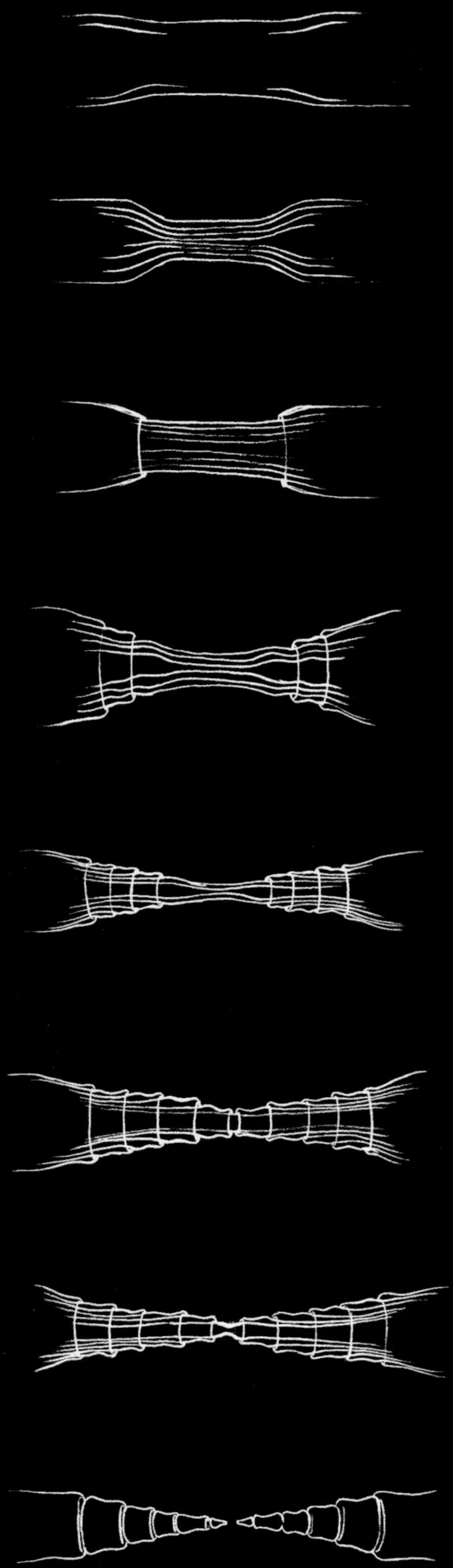

A.

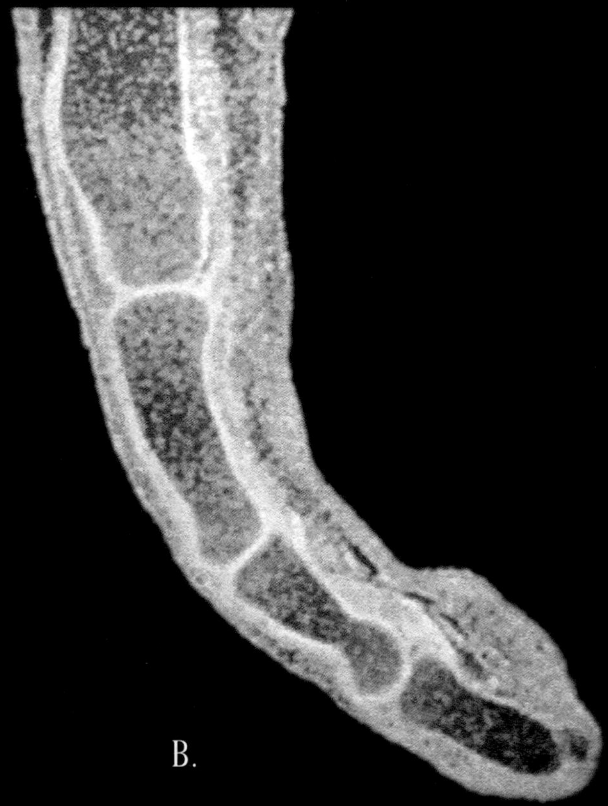

B.

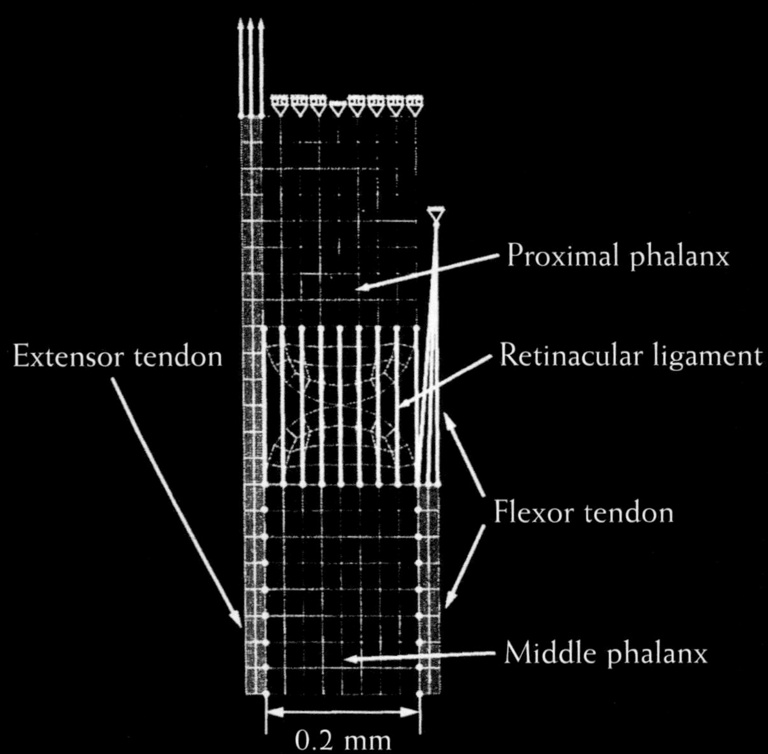

Proximal phalanx
Extensor tendon
Retinacular ligament
Flexor tendon
Middle phalanx
0.2 mm

C.

The centrioles come in pairs (note that plant cells do not have centrioles) that are oriented perpendicular to each other. Albrecht-Buehler (1992a, 1992b, 1994, 1995) speculates that the best possible design for a "cellular eye" would look like a pair of centrioles. This work on the motile behavior of cells revealed that centrioles are able to map the direction of light sources by absorbing and deflecting light of certain wavelengths. Each centriole can map the angle in a plane perpendicular to its axis. Because light is distributed in three dimensions, a second centriole is needed at right angles (Plate 16). Albrecht-Buehler experiments showed that cells were able to detect microscopic near-infrared light sources at a distance. In 1994, Swedish researchers performed a number of computer experiments that demonstrated that the eye will evolve from a sheet of photosensitive cells (Ball 1999). It is also known that the mesothelial cells in the iridocorneal of the eye have an abundance of cilia and centrioles (Wickham 1976). Centrioles also have an important presence in the rhabdomere differentiation of the compound eye in insects (Home 1975), in teleost (al-Adhami 2001) and in the tree shrew retina (Knabe 1997).

The hypothesis made here, indeed with no real evidence, is that internally lined spheres migrate to the exterior, forming the eyes at the anterior, and the genital primordia at the posterior, with which they are morphological homologues.

PLATE 18: *Mitosis*

Chapter 3 Insects, Crustaceans and Other Arthropods

The adult insect body is subdivided into head, thorax and abdomen. Other arthropods; including spiders, scorpions and crustaceans, combine head and thorax. Vertebrates combine abdomen and thorax.

At the earliest observable stages the insect egg appears elongated, segmented, its axis curled, with a band of transverse grooves across the midline of the inside curve called the "germ band." Polymerization of the membrane protein along the fault lines can generate visible lines of force. The oöcyte grows during oögenesis within an unyielding spherical envelope, which can cause circumferential buckling and axial segmentation along the force lines. With increased axial growth, the egg is forced to curl within the constraining sphere. Increased axial growth within the constraining envelope could cause the segments to crumple and compress within a deep furrow, and fuse as flat tubes that cross the midline as the edges of the furrow close over them, pressing them together. With increase of girth, the axis straightens, the midline splits and separates laterally. The formed limbs then emerge and stand forth from the surface, achieving the posture of the arthropod and amphibian limb configuration, an arched tunnel that elevates an axial body above the ground and can manipulate its attitude in all planes (Plate 19).

PLATE 19: *Spider and Scorpion. The egg appears elongated, segmented, its axis curled, with a band of transverse grooves across the midline of the inside curve called the Germ band. Polymerization of the membrane protein along the fault lines can generate visible lines of force. The oöcyte grows within an unyielding spherical envelope, which can cause circumferential buckling and axial segmentation along the force lines. With increased axial growth, the egg is forced to curl within the constraining sphere causing the segments to crumple and compress within a deep furrow, and fuse as flat tubes that cross the midline. The edges of the furrow close over them, pressing them together. With increase of girth, the axis straightens, the midline splits and separates laterally, and the formed limbs emerge and stand forth from the surface, achieving the posture of the arthropod and amphibian limb configuration.*

Metamorphosis

Here is a hypothetical model for the amazing process by which a caterpillar becomes a butterfly. The primordial toroidal structure of the elongated insect egg folds in on itself along a line from pole to pole. In lepidoptera, the in-rolling fold occurs only in the outer part of the torus, the inner sphere experiencing no growth. The inner sphere remains molecular in scale and is relegated to a location on the dorsal midline at the head. Later, all at once, the substance of the caterpillar flows back into the invisible inner sphere, which expands at the expense of the caterpillar to form the new butterfly body (Plates 20, 21).

Butterfly Wings

Insect wings result from a remarkable phenomenon, even by the standards of some of the extreme topological transformations already described. It is as though insects wear a body stocking beneath the outer shell that bears the radial axial pattern of the egg. This is pulled out like a shirttail through a slit at the back of the thorax. The

PLATE 19

whole then fans out displaying the familiar radial pattern of wing veins.

The patterns of butterfly wings are clues to the origin of insect wings. At an early stage, the inner layers of the thorax and the abdomen surfaces split ventrally and then compress dorsally causing the pattern to fold in pleats like a fan. Next, the figures are drawn anteriorly where they reside within the thoracic cavity. Later, the structures retreat posteriorly forcing their way through the posterior dorsal ridge, emerging exteriorly toward the posterior, while they fan out and unfurl (Plate 21).

This pattern originates from the array of thoracic and abdominal segments, which are attached perpendicularly to a lateral stripe that divides the hemispheres of the cell from pole to pole. As the stripes are pulled to the anterior, the segments drag behind subtending an acute angle. The abdominal pair enters the thorax and takes place beside the thorax pair. The thoracic pair lacks the segments on the ventral side as these had previously folded inward forming legs. Thus, the anterior wings have only half the vein pattern of the posterior wings.

The gorgeous patterns of butterfly wings result as a second pattern is superimposed on the vein pattern. This is the result of tinting of the compressed wing membranes within the three segments of the thorax where their folded edges contact the pigmented interior of the segments. These open out in a fan-shaped example of the tie-dye effect.

PLATE 20: *Types of Larvae*

PLATE 21: *Metamorphosis: In Lepidoptera. The in-rolling fold occurs only in the outer sphere of the torus, the inner sphere experiencing no growth. The inner sphere remains molecular in scale and is relegated to a location on the dorsal midline at the head. All at once the substance of the caterpillar flows into the invisible inner sphere, which expands at the expense of the caterpillar to form the new butterfly body. The patterns of butterfly wings are clues to the origin of insect wings. At an early stage, the inner layer of the thorax and the abdomen surfaces split ventrally and then compress dorsally causing the pattern to fold in pleats like a fan. Next, the figures are drawn anteriorly where they reside within the thoracic cavity. Later, the structures retreat posteriorly, forcing their way through the posterior dorsal ridge, emerging exteriorly toward the posterior, while they fan out and unfurl.*

Plate 20

Chapter 4 The Anatomy of Flowers and Fruits

The concordance of flower and fruit parts to the hypothetical torus shapes of the egg cell is direct. Flowers and fruits are shaped as enlarged versions of this figure produced by the change in the size proportions of its parts. The exosphere generates the petals and stamens, the endosphere, the ovary. The top funnel vortex forms enclosed folds, originating the anthers and stigmata. The bottom funnel vortex forms the sepals. The formation of the flower occurs as the direction of toroidal streaming is reversed, causing the vortex to project axially, forming the style and stigmas (Plate 22). The exosphere surface retreats, resulting in petal formation.

Compound flowers, notably the sunflowers, exhibit the intriguing geometry of phyllotaxis in the spiral patterns that are generated in the efflorescent centers of this class of flowers. The pattern is formed of counter-rotating logarithmic spirals, which generate Fibonacci numbers and the proportions of the golden section, lending a certain alchemical quality to the extensive study of the causes of phyllotaxis (Plate 23).

A mechanical explanation of the phenomenon follows: Phyllotaxis is the simple topological-geometric result of the separation of the fluid stream as it enters the vortex, dividing automatically into three, four or five separate vortices arranged around the axis. The release of the compressed pattern of these tubular sheets as they leave the vortex causes them to overlap, forming a wave interference pattern of the type made by the wake of a boat. The ovary is subdivided by the pattern, each section of which generates a flower and a seed. The resulting patterns however magnificent are gratuitous. Fibonacci numbers and golden section phenomena appear everywhere, simply as a property of numbers, and do not necessarily have anything to do with the origin of the system that generates them (Plate 24).

PLATE 22: *The Hypothetical Toroidal Origin of Flowers*

PLATE 22

PLATE 23: *Phyllotaxis is the simple topological-geometric result of the separation of the fluid stream as it enters the vortex, dividing automatically into three, four or five separate vortices arranged around the axis. The release of the compressed pattern of these tubular sheets as they leave the vortex causes them to overlap, forming a wave pattern of the type made by the wake of a boat. The ovary is subdivided by the pattern, each section of which generates a flower and a seed. The resulting patterns however magnificent are gratuitous.*

Plate 23

PLATE 24: *Flowers and fruits are shaped as the exosphere generates the petals and stamens, the endosphere and the ovary. The top funnel vortex forms enclosed folds, originating the anthers and stigmata. The bottom funnel vortex forms the sepals.*

Chapter 5 Toroidal Membranes

There is no escaping the vortex. It is there when you stir your cup of coffee, and when air and fuel mix in your car's engine. It's at work in the giant swirling storm that is Jupiter's great red spot, and in the toilet bowl after you flush it. There is a reason for the ubiquity of the vortex. Given the slightest excuse, any fluid will swirl into a vortex as it is moved. You can't turn around without hitting a vortex or creating one. They're everywhere. (Gorder 2004; Mohseni 2001)

Recent evidence for the toroidal cell hypothesis comes from the work on biomembranes from Lipowsky.

Biomembranes are thin highly flexible sheets composed of lipids and other amphiphilic molecules (molecules having both a water soluble end and a non water soluble end). Bilayer membranes have classical modes of deformation including stretching, bending, shearing and slipping.

In aqueous solutions, low concentrations of lipid molecules will spontaneously form biomembranes, then auto assemble as tubules and, then with swelling, turn into vesicles, coming together in such a way that they have no edges and are closed...on the micrometer scale, the behavior of membranes and vesicles can be understood, to a large extent, in terms of a few parameters. Some of the parameters, which are crucial for the membrane morphology, are the vesicle volume, bending rigidity and the preferred or 'spontaneous' curvature of the membrane. The shape of the vesicles is governed by this curvature and bending elasticity. Other local control mechanism for the curvature of these vesicles is provided by the molecules attached to the membranes which act as hydrophobic anchors.

They have closed surfaces without edges.

They are highly flexible and easily adapt their shape to external perturbations.

In spite of their flexibility, they are rather robust and keep their structural integrity even for strong deformations.

Their basic building block is a bilayer of lipid molecule.

These lipid bilayers are maintained in a fluid state, which is the main mechanism underlying their enormous flexibility. (Lipowsky 1998)

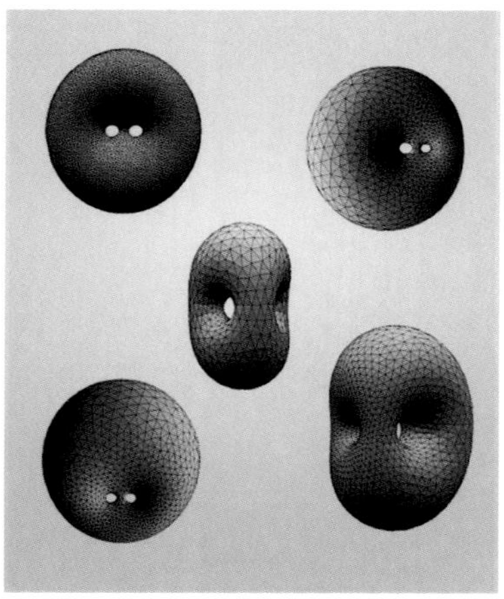

PLATE 25: *The Classification of Toroidal Surfaces* (Jülicher 1993).

PLATE 26: *An Artificially Generated Axisymmetric Sickle-Shaped Torus* (Jülicher 1993).

Being highly flexible, these bilayer membranes can undergo shape changes such as the formation of small sphere-like buds or concave vesicle shapes. A well-known example is the crenated shape of echinocytes of a human red blood cell, which can connect at ends to form a torus. This flexibility is due to the fluidity of the molecules with the membrane.

To a large extent, the fascinating morphology of biological membranes is due to their fluidity, in other words, to the ability of the molecules within them to move laterally along the membranes. This property becomes apparent when studying lipid bilayers that contain only one or a few molecular components. Indeed, in their fluid state, even relatively simple lipid membranes exhibit a rather complex morphology. (Jülicher 1993)

Bilayer membranes will adjust the lipid composition of their membranes in such a way that they remain in a fluid state irrespective of temperature and other conditions. Prokaryotic cells achieve this by increasing the number of unsaturated double bonds within the hydrocarbons of the lipids. These unsaturated bonds act as defects within the bilayer that prevent freezing. (Lipowsky 1998)

The increase in cholesterol increases its fluidity and at the same time affects membrane lysis or rupture tension, thus biomembranes will differ in their ability to stretch by as much as 5% before the membrane ruptures. (Lipowsky 1999)

On the molecular scale, this fluidity is necessary for the proper functioning of membrane proteins. On the supramolecular scale, it insures that the biomembrane is highly flexible and can undergo shape changes... therefore both Prokaryotic and Eukaryotic cells adjust the lipid composition of their membranes in such a way that they remain in a fluid state irrespective of the ambient temperature and of other external conditions. (Lipowsky 1998)

These vesicles are described by Jülicher, Seifert and Lipowsky with a classification resulting from the topology of their closed surfaces, that is by "the number of handles or holes" (Jülicher 1993). Plate 25 shows the shape transformation of vesicles both from phase contrast microscopy and calculated from curvature models. These transformations were induced by an increase in temperature that leads to in increase of the surface area of the vesicle. Beside temperature, vesicle topological transformation can be brought about by changes in osmotic conditions (Berndl 1990; Käs 1991; Farge 1992).

These vesicles exhibit a rich variety of shapes and, until recently, their taxonomy many has mainly concentrated on spherical topology, including discocytes, stomatocytes or even budded shapes.... In 1991, however, experimental work by Mutz and Bensimon showed that toroidal vesicles were experimentally observable among the topological transformations. These membrane vesicles include toroids, not only in the shape of the crenated red blood cells as previously seen, but in more complicated toroidal shapes. Based on the works of Canham (1970), Helrich (1973) and Evans (1980), the shape problem for a vesicle is the minimization of an elastic bending energy for a closed surface at a given area (i.e. fixed number of molecules) and volume (i.e. no permeation through the membrane). Under some circumstances... the latter constraint can be however temporarily released. For vesicles with torcidal topology, the minimization program has been considered both in mathematics and later in physics. (Fourcade 1992)

These toroidal vesicles formed by lipid bilayers are now seen in a number of laboratories and the toroid structures include one hole (genus-1) and two hole torus structures (genus-2) (Jülicher 1993).

Even though the shapes shown... are quite different from the shape of a sphere, these shapes will have the same topology as a sphere. A different topology is obtained for toroidal vesicles with one or several 'handles'. Toroidal shapes have long been studies by mathematicians but the area and volume constraints as applicable to vesicles have been included only quite recently.... One interesting aspect of toroidal vesicles with two or more handles is the appearance of a new diffusion process, termed conformal diffusion, which was first predicted theoretically by Jülicher, Seifert and Lipowsky (1993). In this process, the vesicle undergoes continuous transformations of the shape.... (Lipowsky 1995)

The toroidal vesicles were first observed using polymerizable phospholipids, but later, Lipowsky and his group were able to observe toroidal vesicles using standard non-polymerizable phospholipids.

Recently toroidal vesicles have been observed experimentally. The first examples were vesicles made of a partially polymerized bilayer. Typically, these vesicles were close to one special axisymmetric circular state. the so-called Clifford torus (think of a donut). Subsequently, toroidal vesicles consisting of fluid membranes have been found. They exhibit a variety of different types of shapes of genus one, two and even higher genus. Among the genus-1 surfaces observed so far, axisymmetric and no-axisymmetric shapes occur. The genus-2 surfaces which are necessarily non-axisymmetric can crudely be classified into two classes, 1) discoid shapes with two holes and ii) shapes consisting of two spherical parts connected by three necks. (Jülicher 1993)

These toroidal vesicles can be manipulated and the axial symmetry broken through osmotic stress, tension-induced adhesion and even shape transformations induced by optical tweezers. Optical tweezers provide localized light traps and can be used to manipulate various objects in the micrometer size range (Lipowsky 1999).

Using conformational transformations of the Clifford torus, Jülicher and his group were able to obtain axisymmetric sickle shaped torus as shown in Plate 26.

The systematic theoretical study of toroidal vesicles of genus-1 started with a minimal model fro symmetric bilayers (without spontaneous curvature) with the volume to area ration, i.e. the reduced volume, as the only parameter.... The non-axisymmetric toroidal shapes of lowest bending energy, as obtained via conformal transformations of the Clifford torus, represent the ground state...indicates that regions of non-axisymmetric shapes exist in the phase diagrams of both models. We now apply the hypothesis that conformal transformations of the various sheets for all regions of the phase diagram. Specifically, we analyze the infinitesimal stability of the axisymmetric sheets with respect to special conformal transformations which can be calculated from the contour line of this shape. This method gives a lower bound on that region of the phase diagram in which the ground state is non-axisymmetric. The idea behind this method is to start with a shape on the sheet of lowest energy shapes and apply a special conformal transformation.... The stability of the original shape now depends on the difference of the energy. (Jülicher 1993)

In a recent article in *Science* magazine, Pochan et al (2004) were able to produce toroids from micelles using blocked dissimilar polymers, joined together. Triblocked copolymers can form toroidal geometry. "The key to forming this unusual morphology was the addition of a positively charged divalent organic counterion. This produces a self-attraction of cylindrical micelles that leads to toroid formation, a mechanism akin to the toroidal bundling of semi-flexible charged biopolymers such as DNA." These toroid services had a high degree of control and uniformity (Pochan 2004). "The micelles are formed by a combination of cylindrical micelle curling and end-to-end collapse of cylindrical and spherical micelles. The size of the toroidal micelles can be influenced with the concentration of diamine counterion."

If the radius from the center of the hole to the center of the torus tube is c, and the radius of the tube is a, then the equation in Cartesian coordinates for a torus symmetric about the z-axis is:

$$(c-\sqrt{x^2 + y^2})^2 + z^2 = a^2$$

where:

$$x = (c + a \cos v) \cos u$$
$$y = (c + a \cos v) \sin u$$
$$z = a \sin v$$

Chapter 6 Autogenesis

Autogenesis is a convenient term for the theory of biological self-organization. It accounts for the construction of the body and its change over time in terms of geometry and physics; that is, the shaping of the body by natural patterns that occur as forces act on the growing embryo, rather than being directed by an inherited roadmap. The theory constitutes the algorithmic, geometric sequence of changes of a model egg, which predicts with accuracy the body form, leaving the details to the genes.

Laymen to biology are surprised to learn that few developmental and evolutionary biologists any longer believe that natural selection plays any significant role in the creation of form, but is rather relegated to microevolution, having to do with such as the shapes of the beaks of finches.

"No Need to Spell It Out" identified a viewpoint by Adrian Cho in a review in the journal *Science* (Feb 6, 2004) entitled "Life's Patterns," which characterizes the idea that "rather than being specified by the genes, elaborate biological structures may arise through simple dynamical mechanisms." The review brings together the work of a dozen developmental biologists who believe that form is not entirely directed by the genome. Biologists now speak of genes as switches that engage some other mechanism, which actually forms the body. Many researchers assume that form is shaped by the self-organization of the growing cells in specific patterns, in response to predictable, immediate, physical forces. This goes against half a century of biological dogma that form is encoded in the genome, and that evolution occurs by the natural selection of random mutants resulting from accidental changes in the code.

From the seventeenth through the nineteenth century, biology consisted of the pursuit of the secret of life form for which Goethe coined the term "morphology" and for which he may indeed have been ready to sell his soul. The race to complete a model of morphogenesis was interrupted in 1859 by Darwin's publication of a hypothetical phenomenon that could account for the generation of anything, much like divine creation. If a breeder can make a new dog in several generations, think how many test trials nature can run in millions of years before selecting the perfect shape of every organ.

In 1953 the DNA molecule was discovered to be so vast that it was thought to contain a code for the entire blueprint of the body (Watson & Crick 1953). Since

PLATE 27: Various Invertebrate Phyla

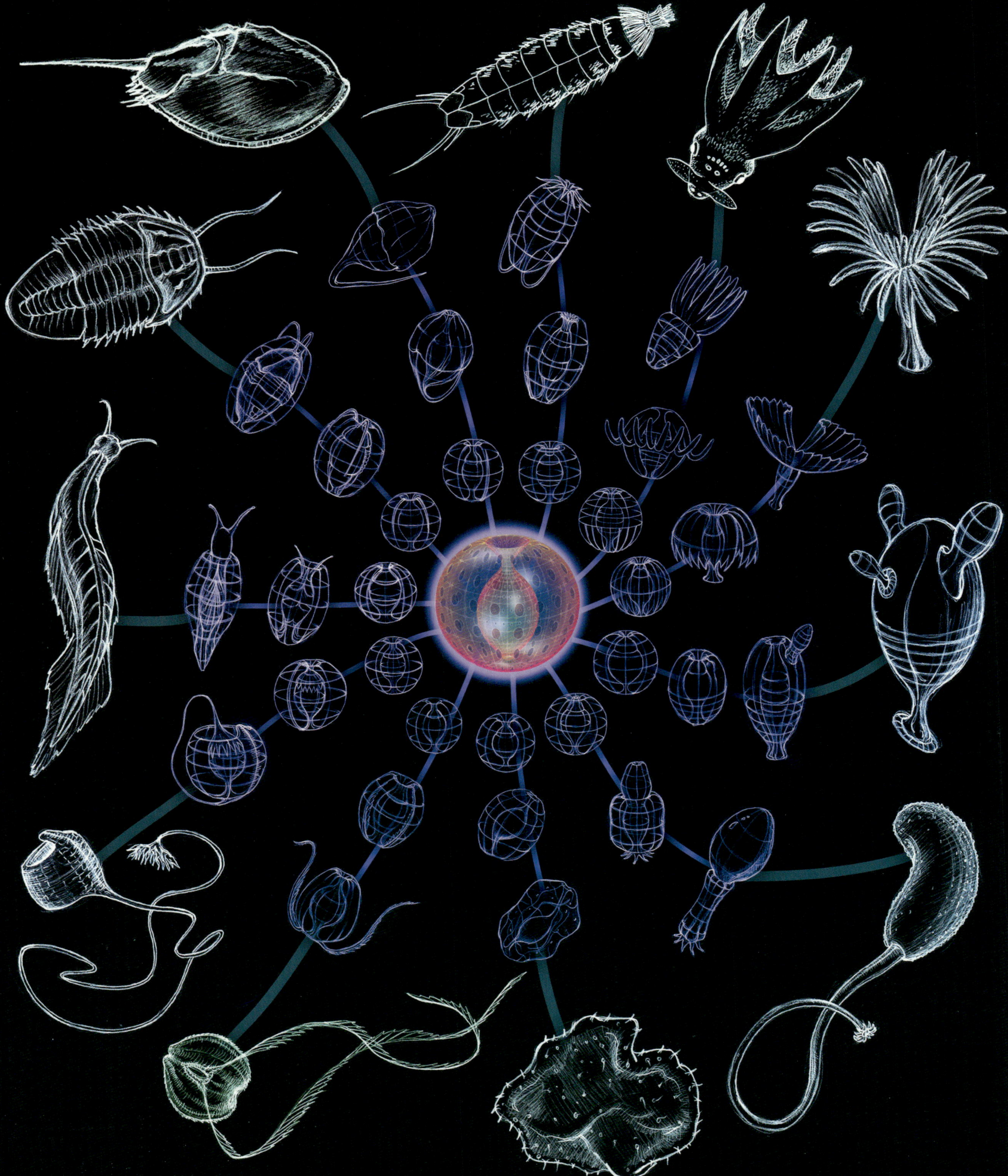

Plate 27

then an immense effort was expended to crack a code for form in the genome. In 1980 Edward Lewis, Christiane Nusslein-Volhard and Eric Wieschaus identified the genes that control segmentation in fruit flies, a research effort for which they later won the Nobel Prize. However, fifty years of intense effort have failed to produce any evidence of the generation of actual form by genes. Yet thousands of researchers continue the attempt to associate genes with specific structures. By now many, if not most developmental biologists, conclude that the code for the construction of the body form lies elsewhere. The sequencing of the genomes of many organisms is now complete. However, it is not a blueprint as many once thought. It is not even a plan. Changing the blueprint might alter genotype and might alter scale, but basic body form, the phenotype, remains the same. Decoding the human genome provides the panacea of genetic engineering, but tells nothing about how the body is constructed. Still, hundreds of millions of dollars in research in developmental genetics is spent yearly seeking a nonexistent code for form in the genome.

The review in *Science*, cited above, identifies a communality of thought, which it dubs "Dynamical Mechanisms." Ironically the concept that unites this band of researchers was once the predominant belief in theoretical biology before the current era of Darwin and DNA. The newly identified dissenters maintain that there is a wholesale misunderstanding of what the genes do and how evolution works (Milton 1998; McIver 1988; Stewart 2003). The emerging consensus is that the phenotype is generated by the self-organization of the elements of the developing embryo by a process or processes unknown.

Darwinism has instructed us to see ourselves as randomly created over time to perfectly suit our environment, and that there is no logic to the form of our bodies except functional design. But the Faustian secret of the morphologist is that our bodies share a single, natural, universal pattern, which occurs in many varieties, like snowflakes. The perceived perfection of design of our bodies originates in the actual perfection of spatial organization that occurs when molecules stack. Magnified, the process can produce the marvels of intricately organized symmetry in flowers, butterfly wings, snakeskins, snail shells, or perfect calcite crystals the size of a telephone booth. Biodiversity is chiefly interesting for the clues it provides in its generating principles. Otherwise, to use a famous analogy, it is more like bird watching than ornithology.

What, suggested D'Arcy Thompson, could be more unnecessary than invoking millions of years of selective fine tuning to explain the shape of a horn or a shell when one could propose a very simple growth law based on proximate physical causes to account for it. The sabre-like sweep of an ibex horn does not have to be selected from a gallery of bizarre and ornate alternative horn shapes: we can merely assume that the horn grows at a progressively slower rate from one side of the circumference to the other, and hey presto—you have an arc... Thompson's point was that it need not have been selected—it was inevitable. Either the horn grew at the same rate all around the circumference, in which case it was straight, or there was this imbalance from one side to the other, giving a smooth curve... the shapes of amoeba say, can no more be regarded as 'selected for' then can the spherical form of a water droplet; rather they are dictated by physical and chemical forces. (Ball 1999)

PLATE 28: Hydra

In his magnum opus, *On Growth and Form,* the great morphologist D'Arcy

Thompson sought to rationalize nature in a logic described by Stephen Jay Gould as a combination of Pythagoras and Isaac Newton.

Far from being randomly formed, our bodies are the realization of geometry, the fundamental order implicit in the partitioning of space. Thus, any small mass of protoplasm can inevitably reproduce one of the many phyla including a large-brained vertebrate with a ventral face (Plates 27, 28). The understanding of organic form as independent of environmental change increases by many orders of magnitude the likelihood of the existence of complex life and intelligence on many of the myriad earthlike planets.

The proposed solution to the riddle of morphogenesis by autogenesis is based on various disciplines, including toroidal surface topology, colloid chemistry, geometry, physics, mechanics, self-organized criticality, and particularly a subject called instability and collapse. This follows the long tradition of morphology that sees biology as a branch of physics.

With form come mechanisms for movements that automatically create controlled locomotion. The animal, acting by means of the movements it inherits, assumes a behavior pattern. It does what it can with what it has (Plate 28). If it survives, it appears it was perfectly designed for whatever it ends up doing. But that is an example of teleological fallacy, to imbue purpose where there is none. Nature has no purpose, direction or goals, except perhaps to complete the geometric patterns it begins.

All the animal phyla, which we now count at around three dozen, dramatically appeared, ready made, around five hundred million years ago in the so-called Cambrian explosion. No new ones have appeared since. This partitioning of the morphological niches may be compared to the geometric systems of crystals, which coincidentally are roughly of the same number. On remarking on the regularity of many living forms, especially the small and less developed, scholars for centuries have sought an algorithmic system, a set of construction rules, capable of generating these forms, comparable to the simple system governing the stacking of atoms and ions as they precipitate from solution or pass from liquid to solid.

The hypothetical sequence of topological events takes place in the submicroscopic realm where surface tension rules the movement of a transparent network of molecules that subdivide the cell membrane in invisible, self-assembled patterns. These will become evident only later upon the subdivision of the patterned membrane into a mosaic of individual cells. The pattern emerges like the image on an exposed photographic plate in the developing tray. The separate proliferation of these cells, called blastomeres, causes the enlargement of the growing embryo, while preserving the original configuration of the patterning in the egg, and allowing for differentiation by varying the rates of growth of the parts.

Ontogeny, the development and growth of the individual is two separate processes: the establishment of the embryonic form and its growth into adult size. The adult is a mechanical enlargement of the mature egg by the mosaic subdivision of the surface into separate cells. Each mosaic subdivision becomes a center of proliferation, resulting in the enlargement of the form while preserving the original shape.

The main observable event in animal embryogenesis is gastrulation, known to every high-school biology student as the cause of the body germ layers as the growing cells enter their own interior. When the empatterned egg cell has grown to a thousand times the normal cell size, it is suddenly reduced by subdivision to a ball of a thousand normal-size cells, whereupon gastrulation is begun. The continued expansion of the membrane, now a sheet of cells, is forced to enter the interior, either along a meridian on the ventral side, or through a small aperture called the blastopore, effectively turning itself inside out. This causes considerable plastic deformation of the elements of the structure as they pass through the blastopore. Gastrulation is responsible for directly forming the body parts, differentiating the resulting configurations into separate phyletic groups. Meridian gastrulation produces the arthropods; polar gastrulation the vertebrates. All animals pass through a gastrula state.

It is not birth, marriage or death, but gastrulation, which is truly the most important time in our life. (*Slack* 1983)

Chapter 7 The Crisis in Biology

DNA did not create life; life created DNA.... DNA is a mechanism created by the cell to store information produced by the cell. Once produced by the cell, DNA became a stable place to store structural information about the cell's chemistry, something like the minutes taken by a secretary at a noisy committee meeting. (Commoner 2002)

How does a single fertilized cell morph into an organism containing trillions of cells? Why do organisms as varied as a fly, lobster and human share the same fundamental body plan? How over time did simple forms achieve their present complexity? Is it all in the genes or does the answer lie in an alternative explanation, one that involves physics, mathematics and engineering principles as well as inherited characteristics? These fundamental questions of biology remain generally unanswered.

With the discovery of the structure of the DNA molecule in 1953, the vast majority of biological scientists accepted genetics as the key to understanding the fundamental phenomena of biology, including development and evolution. The genome was called the "blueprint of life" and natural selection of random changes in its structure was the phenomenon that shaped species during evolution. This idea was accepted as dogma by biologists (Behe 1996; Zimmer 2001; Ridley 1999). It became almost an article of faith that determining the exact structure of the genome of humans and other species would reveal how the body is assembled by the genes, and solve the problem of morphogenesis. Never mind inquiries like that of Gabriel Dover (2000) who legitimately raised the question: Why 173 pairs of legs on a centipede? Did the centipede arrive at 173 pairs of legs by mutation and natural selection? Are 173 pairs of legs an adaptation to a special kind of environment? Or maybe it is 15 pairs of legs that have adapted to another environment? Do 16 pairs of legs give higher survival levels than 15? And how many accidents of birth are necessary to get 173 pairs of legs?

Rather than launching biology into a new era of understanding, the complete decoding of the human genome announced in December 2000 precipitated a crisis in biology. Long thought to comprise over a hundred thousand genes, the human genome was found to contain merely thirty thousand. The decoding of the mouse genome in December 2002, also numbered thirty thousand, and of these, only three hundred differed from the human genes, a mere 1%. This small number of genes and lack of species diversity in the DNA suggests an insufficiency of information to distinguish the body form of man from mouse (Plate 29).

PLATE 29: The Genome

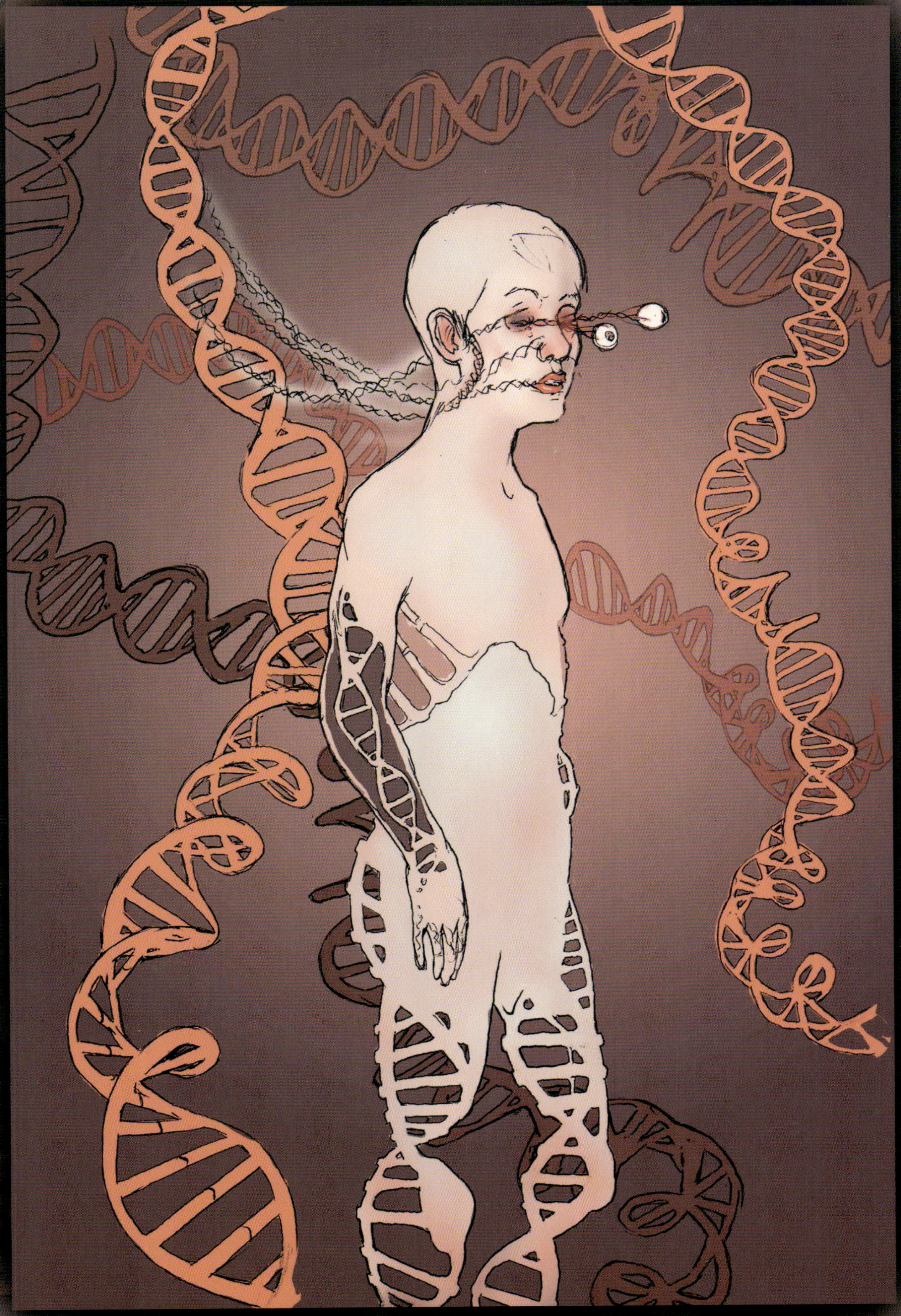

Plate 29

Even before the genome was fully decoded, cracks in the dogma began appearing. As Werner Muller put it in the 1996 edition of *Developmental Biology*, a standard college text:

It is commonly stated that the genome incorporates a Bauplan, an architectural plan, or blueprint of the body. Actually this is not the case; the genome is not a sketch or design of the finished body. The informational capacity of DNA is simply too low to store the blueprint of the final pattern of the organism. For example, the detailed design of the hundred trillion to one quadrillion synaptic contacts of the brain alone would greatly exceed the genomic memory.... We do not understand in detail how a developing organism is created on the basis of such minimal information, or how many organisms are able to regenerate lost structures.

Richard Lewontin, the noted evolutionary biologist, was more caustic. In his preface to Susan Oyama's *The Ontogeny of Information* (2000), he says:

Already in 1982, at the Cambridge University symposium commemorating the one hundredth anniversary of Darwin's death, Sydney Brenner in his keynote address had declared that given the complete DNA sequence of an organism and a large enough computer it would soon be possible to compute the organism. By 1985, the genetic boa constrictor had totally enfolded embryology in its helical coils. Now, nearly fifteen years later, the helpless victim has disappeared down the molecular maw and is slowly being digested. In the end, of course, the result will be what is to be expected from such a process: a slight enlargement of the body of the consumer and the production of a large amount of fecal matter.

The genome has shed little light in the study of morphogenesis. In the December 6, 2002 issue of *Science*, Ray Keller of the University of Virginia began the article "Shaping the Vertebrate Body Plan by Polarized Embryonic Cell Movements" with the admission: "How the body plan is shaped from a cohesive aggregate of individual cells during embryogenesis is an enduring mystery."

The standard body is, of course, produced by the proteins fabricated under the direction of the genes (Plate 29). But now the genome has come to be seen as the determinant of the difference between man and mouse by controlling the relative size of the component parts, by doling out their component material in the right quantity at the right time.

So great has been the success of biology in medicine, especially genetic engineering, that it is amazing to realize biology is a branch of science that lacks a coherent fundamental theory. The difference between chemistry and biology is that in biology none of the data fits. There is no confirmation that the genome is the code for phyletic form. Biology is based on the genetic theory of inheritance, that form lies encoded in the DNA molecule, or genome, which guides the formation of the embryo, occasionally mutating and producing new forms, the "most fit" of which manage to survive. Thirty years ago it was discovered that certain genes control the making of certain organs, and if a particular gene were removed, no wing would grow. It was believed that with enough experimental data the genetic code could be cracked, revealing how every bone and every organ is shaped. This has never happened. Many writers in this field now declare that there is no code and that the embryo is formed by a completely different phenomenon.

Stewart says in his book, *Life's Other Secret* (1998):

Genes may tell the plant or animal which plant or animal to be. That is they select a particular assortment of patterns from the physical grab bag. The patterns themselves, however, are provided by mathematical rules and unless we understand those rules and what mathematics they conceal, the role of the genes will forever remain mysterious.... We *know a lot of DNA, especially how it makes proteins, but we know far less about how those proteins and other DNA products are marshaled together to create a complete organism. It's frustrating. In fact, the problem of biological development is one of the biggest challenges facing science. How do organisms regulate and control their own growth patterns? What defines an animal's body plan? How is this form transformed from the DNA drawing board to the developmental assembly line? How does the living material know where and what to grow? The solutions for these questions involve chemistry, biology, physics and mathematics—not just one of them.*

Edelman (1988) agrees, saying that genes do not store the information on the position in space or time of a given cell during the development of an individual animal and that there is a great deal of variation in cellular activities during morphogenesis among genetically identical individuals.

The cover of *Harpers Magazine*, February 2002, proclaims in huge, black type, *The wonders of genetic science proceed from the premise that the DNA double helix is the sole agent of inheritance in all living things. The premise is false. Barry Commoner takes down the Genome Project.* Commoner's article "Unraveling the DNA Myth, the Spurious Foundation of Genetic Engineering" makes the case that DNA is only a mechanism created by the cell to store information. *Because of their commitment to an obsolete theory, most molecular biologists operate under the assumption that DNA is the secret of life, whereas careful observation of the hierarchy of living processes strongly suggests that it is the other way around.* The quest for the key to living form by decoding the genome is futile, because there is no code for form.

Brian Goodwin and Richard Sole (2000) write in *Signs of Life, How Complexity Pervades Biology*:

We have had the organism [explained] as the expression of a vital principle, as a machine, as a complex chemical network, as a result of natural selection. None of the attempts to characterize life is wrong. Each gives a distinct insight into the nature of organisms, but each is also limited. The latest version sees the secret of life in the DNA coiled at the heart of every cell, organizing the dynamic activities of an organism like a conductor bringing coherent form from the orchestra. Now that we have looked deeply into this secret we see that it too is dissolving, or rather, exploding before our eyes. We are overwhelmed by what appears to be the sheer complexity of the information in the DNA and the problem of making sense of it. Information has meaning only within a context, and the living context still evades us.

Goodwin (1994) further writes:

There is no genetic program for development, no program that guides the system through its morphogenetic transitions.... The genes set the parameter values meaning that they produce component parts of the system within a range of values. The morphological transitions then are consequences of the cycle of dynamics generating geometry and geometry modifying dynamics.... In the life cycle of Acetabularia acetabulum, *there is a whorl formation as the fruiting disc opens, not unlike a mushroom cap. It has no Darwinian explanation, no purpose. It is purely a mechanical function of changes in calcium gradient...it is a form created not by a genetic program but due to the mechanical and chemical response of a sheet of cells to a change in chemical gradient. Eye formation has evolved independently, over forty times, because there is a morphogenetic attractor that specifies form. Eyes are the product of high probability spatial transformations of developing tissues. This is very different from the neo-Darwinian position, which states that organisms generate highly improbable structures, like the eye, that persist because they are useful.*

Harvard evolutionary biologist Richard Lewontin (2000) states:

One... problem that nineteenth-century biology took to be the major challenge for a mechanistic science of life, [is] the problem of the development of form. A great deal is known about the genes that encode various chemical signals in development, and about how the network of signals in development is hooked up, but we do not have the faintest idea about how all of this is turned into the shape of my nose. We do not even know how to ask the question in a useful way, although some interesting models have been made of how chemical signals might affect the arrangements of cells.... (Plate29)

The failure of Darwinism and genomics is the theme of many of the emerging biological essays (Johnson 1991; Denton 1986; Sole 2000; Steele 1998). Richard Lewontin in his preface to Susan Oyama's, *The Ontogeny of Information* (2000), says:

The first edition of The Ontogeny of Information, *1985, was instigated by a growing impoverishment of explanation and understanding of how organisms come to be. As a consequence of a long history of discovery in genetics and especially in molecular genetics, together with a virtually complete stagnation of the classical science of experimental embryology, explanations of how organisms grow and differentiate to produce their characteristic forms and properties had become framed entirely in terms of the "turning on" and "turning off" of genes.*

This same book on its jacket quotes Stephen Jay Gould saying, "Lewontin is the smartest man I ever met." This is no little compliment for a biologist questioning the stagnation of biological theory.

The genotypic features, those features that are different in all individuals, are encoded in the genes. However, nature is able to produce the structure of the standard phyletic body with no directly inherited information, driven by the forces of the gravity-less world of smallness, where surface tension, geometry and physics rule. Living form is indirectly implied by the physical characteristics of protoplasm, much as the cubical shape of a table salt crystal is implicit in the electromagnetic profile of the sodium and chlorine atoms. Inorganic crystals gain their simple geometric forms by the enlargement of configurations, which arise in the perfect geometry of the molecular realm, where the crystal begins forming like a pyramid of oranges. Similarly, organic nature may preserve in growth forms that originate by molecular self-organization shaped by simple geometry.

The wonders of genetic science are all founded on the discovery of the DNA double helix—by Francis Crick and James Watson in 1953—and they proceed from the premise that this molecular structure is the exclusive agent of inheritance in all living things: in the kingdom of molecular genetics, the DNA gene is absolute monarch. Known to molecular biologists as the "central dogma," the premise assumes that an organism's genome—its total complement of DNA genes—should fully account for its characteristic assemblage of inherited traits. The premise unhappily is false. Tested between 1990 and 2001 in one of the largest and most highly publicized scientific undertakings of our time, the Human Genome Project, the theory collapsed under the weight of fact. There are far too few human genes to account for the complexity of our inherited traits...in short, the most dramatic achievement to date of the $3 billion Human Genome Project is the refutation of its own scientific rationale. (Commoner 2002)

Chapter 8 Genes: Blueprints or Switches

There is a naiveté about genetic determinism in both evolution and development that signifies intellectual laziness at best and shameless ignorance at worst when confronted with issues of massive complexity. (Gabriel Dover 2000)

In 1980 Cooke asked, "Shouldn't we do well at this state to be flexible, rather than succumb to the current tendency for each aspect or event in an organism's life that attracts our interest promptly to become the express responsibility of a gene?" Cooke was restating a much older concern of Guyénot in 1929, when he claimed that "genes were not 'determinants' representing the one or the other part of the body; rather, they are modifiers of the developmental processes, intervening in this or that cellular functioning, hence in these or those morphological outcomes."

Many authors now question the role of genes in phyletic form. Although there is presently an actual catalogue of all the genes in organisms such as *Caenorhabditis elegans*, *Drosophila melanogaster* and *Arabidopsis thaliana*, the understanding of genetics has not resolved the issues relating genotype to phenotype (Larsen 2003). We cannot relate the genome to the determination of morphology and form.

Organisms are not just genes and the proteins made from them. They are sugars, lipids, minerals, hormones and inorganic molecules. Cells possess surface tensions, electrical charges and colloidal viscosity. None of these substances are encoded in the DNA and gene hunting will not tell us much about these structures. (Ball 1999)

What is generic within the developing embryo and what is genetic? Newman and Comper (1990) believe that genetic actions simply complement the generic actions of tissues. They define generic actions as those mechanisms and physical processes that are broadly applicable to living and nonliving systems. These include adhesion, surface tension, gravitational effects, viscosity, phase separation, coupling and phase-diffusion. All of these may generate morphogenetic arrangements of cytoplasmic tissue and extracellular matrix components. Similarly, these same authors look at the major types of gastrulation including epiboly, involution and delamination as a result of physical forces. Wilson, Cretekos and Helde (1995) suggest that the process of epiboly may simply be a passive, generic response to the shape and changes in the blastoderm prior to gastrulation.

PLATE 30: *Changes in the Hox genes can cause the distal part of a* Drosophila *antenna to become transformed into a leg (Duncan 1998).*

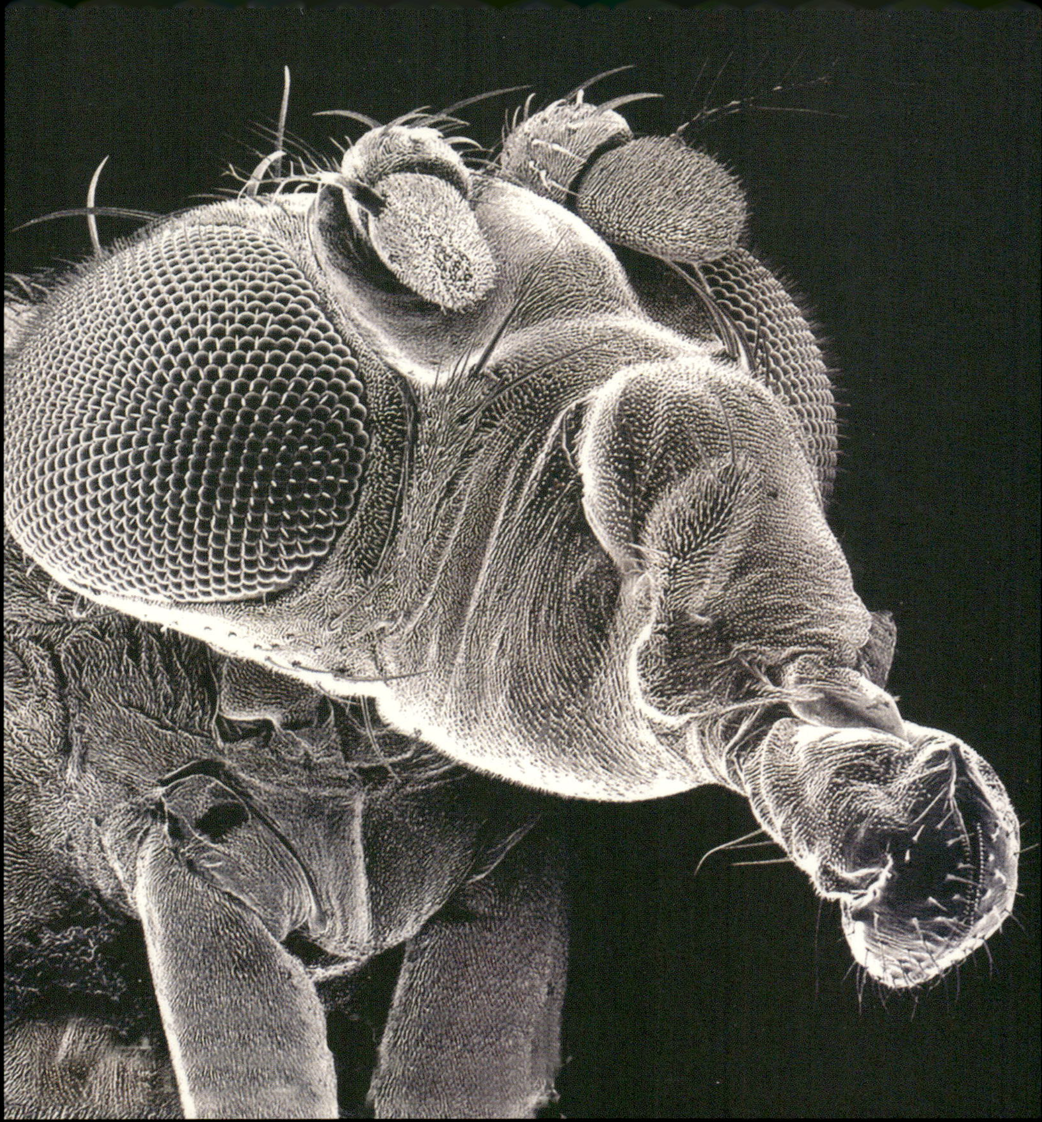

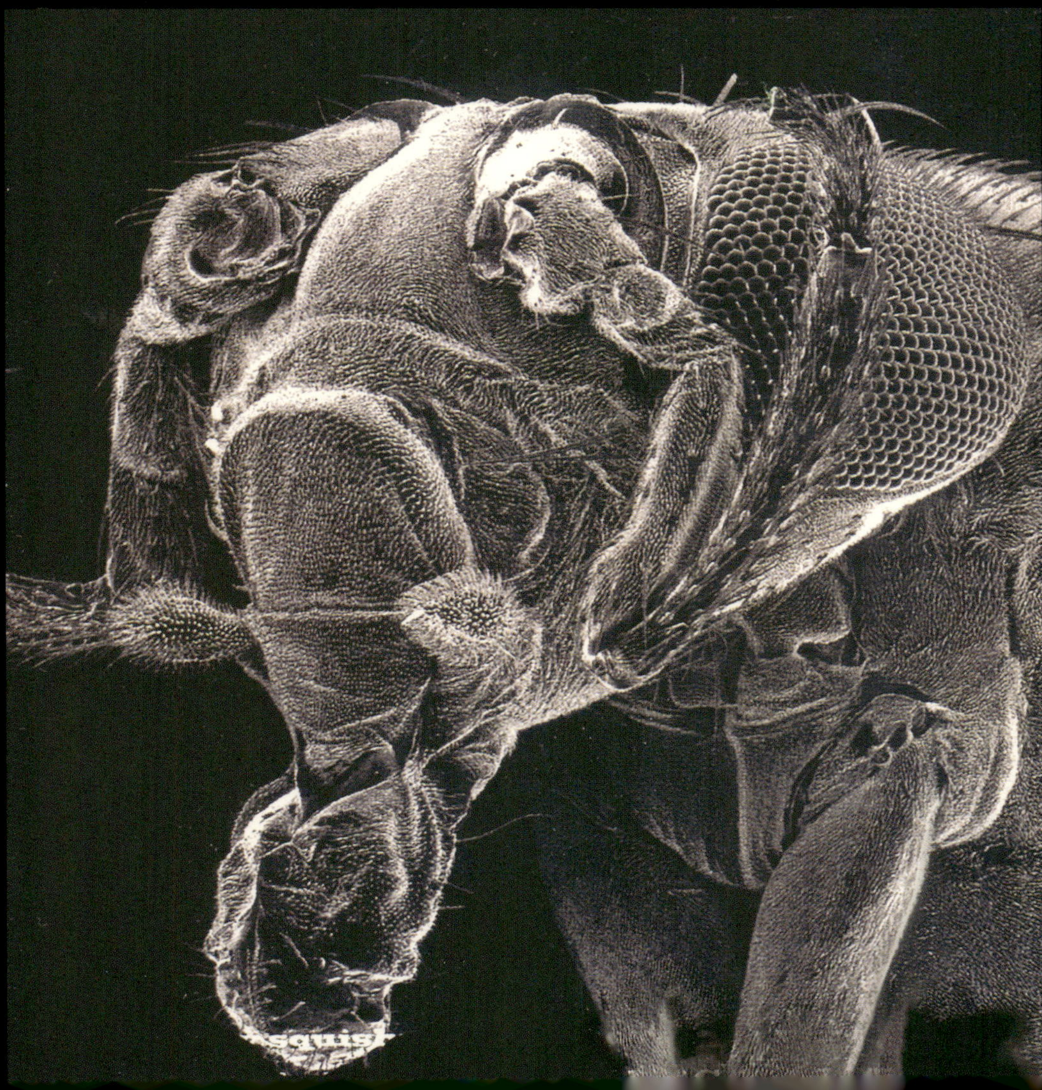

PLATE 30

But if many actions of tissues are generic, one must ask the questions, Are there developmental genes? Can a single change in a gene create a leg for an antenna in a fly? (Plate 30) Minelli (2003) answers "Yes, in the sense that patterns of expressions of many genes are strictly limited to and correlated with specific times and events of development; "Yes," in so far as mutations in these genes may critically and conspicuously alter the normal course of development; but "No" if we take any of them as directly responsible for the origin of an organ or the shaping of the body." Genes do encode transcription facts and components of intercellular signaling but these genes are much the same in animals with very different body plans (Watson 1991a; Davidson 2001). Minelli guesses that these genes have been conserved because of their generic ability to stabilize form, not because of any specific role in generating a particular body plan; for instance, genes may help in stabilizing or maintaining a given spatial arrangement. He quotes Budd (1999) who said, "It is form that captures genes more than it is genes that create form."

Jeffrey Swartz in his book *Sudden Origins* (1999) discusses this as a concept of genetic homeostasis: "Natural selection acts to maintain stability." This concept was first expressed by Williams (1966) in *Adaptation and Natural Selection: A Critique of Some Current Evolutionary Thought*. He wrote that natural selection and mutation could not alone explain the origin of a species. It was his opinion that mutation interrupts the stability that selection tries to maintain. Mutation alone may not call for all the species variations. "Point mutations, which seem to occur most frequently in the third base of a triplet codon in the DNA sequence encoding a specific molecule, tend to have no expressed effect on the molecule's structure and its function in the organism. Consequently a point mutation is essentially silent" (Swartz 1999). Gould and Eldredge (1977) also argued the inconsequential effect of point mutations:

We do not see how point mutations in structural genes can lead, even by gradual accumulation to new morphological designs.... The near identity of humans and chimps for structural genes, and the evidence of major regulatory change indicated by human neoteny provides an important confirmation.

Davidson (2001) in *Genomic Regulatory Systems: Development and Evolution* also believes that there are no genes specific to a given body plan. He says, "There are no insect genes, no sea urchin genes, and no vertebrate genes. Furthermore, differences between a beetle and a frog are not a matter of building stones, nor are they a matter of the kind of paper and ink used to draw a blueprint of their respective body plans." Davidson is not alone in this view. Wagner, Chiu and Laubichler (2001) conclude that, "Evidence at hand does not demonstrate that any particular gene is instrumental in the origin of major characters, such as insect wings." One gene can seem to control the development of many forms. For instance two homeobox genes, *nubbin* and *apterous*, are expressed in similar ways both in the wings of a fruit fly and in the limbs of *Artemia*, a shrimp-like crustacean (Swartz 1999). Sometimes something as simple as the stimulation of cell growth is enough to produce changes in body form and pattern. Swartz discusses that the major reason the same two kinds of cells can produce such different structures as hair and feathers is nothing more than the structural process of invagination or evagination. If a growing mass of cells invaginate, the result will be hair in a follicle or a skin gland. If the group of cells evaginate, the result will be scales or feathers. All of these structures occur because of the protein keratin, a protein common throughout the animal kingdom; the difference of invagination or evagination being due to the mechanical and elastic properties of the cells and the uptake of water (Swartz 1999; Oster 1982). Oster and Alberch propose that the key to this hydration is the accumulation of large macromolecules, named acidic mucopolysaccharides, a key molecule in living forms and one that attracts water molecules (1982).

Many call the current interest in developmental genetics "default morphology," meaning that to attribute formation of a given body part to the specific activity of one or a few genes is unwarranted. "It is incorrect to conclude that a gene must be a major or pivotal one, if a particular process does not work in its absence" (Lawrence 2001). "The gene is ignorant of its role in development" (Dover 2000). Developmental genetics is discovering more and more examples of default morphology manifested when some gene activity is lacking. "One could ideally push gene silencing further and further, until a minimum gene number is found, by which a given level of morphological complexity is still generated or, better, saved and transmitted" (Minelli 2003).

Default morphology in gene control is found in many examples in the literature. In 1984, the homeobox genes were discovered that worked in the development of the fruit fly, *Drosophila* (Plate 31). Later, it was shown that these same homeobox clusters could be found in vertebrates. They eventually were found in chordates, echinoderms, arthropods and nematodes. When Edward Lewis, Christiane Nusslein-Volhard and Eric Wieschaus won the Nobel Prize for the identification of the genes that control segmentation in fruit flies, gene research became the central focus of developmental biology. A gene-centered view of development has dominated the biological

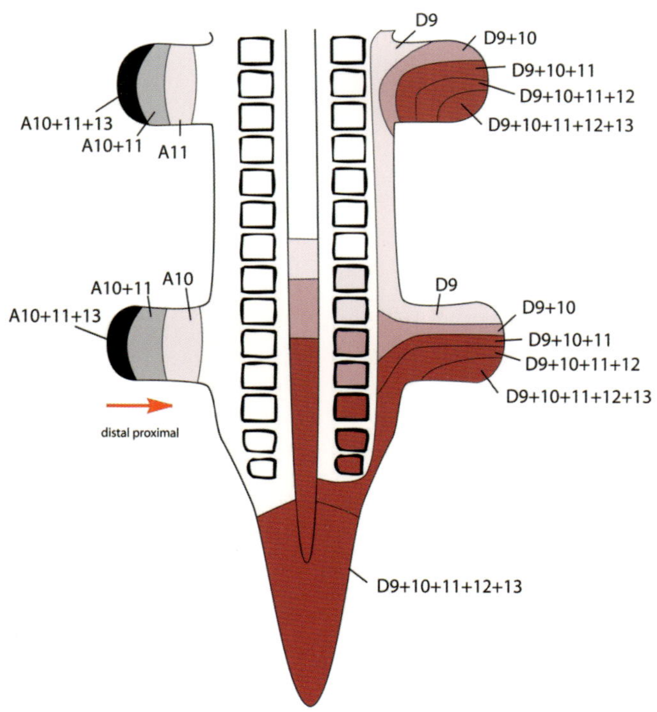

PLATE 31: *Homeobox Gene Expression* (*Müller 1997*)

literature for the last few decades. "Who of us in development biology have not been impressed by changes in the Hox genes that cause the distal part of a *Drosophila* antenna to become transformed into a leg?" (Duncan 1998)

It was Caroll in 1995 who demonstrated that insect-wing formation, although subject to control by Hox genes, is not created by any of them. Wing production can be extended to most trunk segments, but can be repressed in different body segments by still different Hox genes! One might see a leg instead of an antenna in the *Drosophila,* but no gene manipulation will produce a finger coming from the nose of a fly. Casares and Mann (1998) said that the ground state of the ventral appendages in *Drosophila* (antennae, legs, genitalia and analia) is a leg-like appendage consisting of a proximal segment and a distal tarus. For instance the expression of two genes, *proboscipedia* and *sex combs reduced* is required for the specification of sucking mouthparts. In the absence of the activities of these two genes, an antenna forms in place of the proboscis (Percival Smith 1997). In the absence of the activity of the gene *spineless* the distal part of a *Drosophila* antenna is transformed into a leg (Duncan 1998). "Rather than acting as direct controllers of developmental processes, the genes act as selectors, steering development among alternative options" (Nijhout 1990). Like Ball, Nijhout concluded that development had to be considered not just in light of gene expression, but also as a consequence of structural, chemical and physiochemical components. Nijhout also rejected the notion that the genome contains any developmental program. He uses the example of the *bicoid* gene, again in *Drosophila*, whose expression is now deemed essential in the determination of the anterior-posterior axis in the body plan. Nijhout says that what is being overlooked is the fact that the correct gradient of the *bicoid* product is established only if other products in the insect embryo are produced at the same time, allowing structural interactions in the cytoplasm. Ellen Larsen, a prominent biologist in the genetics of *Drosophila*, similarly believes the relatively few changes in cell behavior may be responsible for large morphological changes without unraveling the rest of development (Larsen 2003).

"The role of Hox genes in determining major changes in body architecture has been extensively debated" (Minelli 2003). Hox genes do not play any role in patterning the embryos or larvae of many bilaterians, such as sea urchins and polychaete annelids. In *Caenorhabditis elegans* (a small soil nematode), an essentially complete embryo is produced in the absence of any Hox gene expression (Salser 1994; Van Auken 2000). And in many developing organisms, Hox genes contribute only to the adult form and not to the larval forms (Williamson 2003; Peterson 2000).

The idea of master control genes in development was first introduce by Lewis (1992). "A master control gene would be responsible for a major switch in the expression of a large number of downstream genes" (Minelli 2003). Geneticists now generally believe that genes are switches that instigate a cascade of other activity leading to the formation of segments in the egg and the embryo. The standard paradigm is that the gene creates form, as proteins produced by the gene initiate cell division timed to correctly reproduce the body. According to Evelyn Keller (2000) in *The Century of the Gene*, the notion of a genetic program depends on the common mistake of equating the distinction between genetic and epigenetic with the distinction between program and data. The genetic information present in a zygote is usually thought of as a computer program, but it might be more accurate to consider it as a set of data that needs to be processed by some internal program.

For instance, the *eyeless* (*ey* gene) in *Drosophila*, a homologue of the *Pax6* in *Xenopus*, was long thought to be a master control gene for eye morphogenesis (Halder 1995; Chow 1999). The involvement of *Pax6* homologues was found in squid (Tomarev 1997), amphioxus (Glardon 1998), nemerteans (Kmita-Cunisse 1998) and planarians (Pineda 2000). Developmental geneticists were quick to conclude that the origin of the eye is monophyletic throughout the animal kingdom (Gehring 1999; Gehring 2000). However, later, Wagner (2000) found that the *Pax6* gene initiated the development of needed proteins for the early development of light sensitive epithelia (see possible role of centrioles Chapter 2), a simple precursor of both the camera eye of vertebrates and squids and the compound eye of arthropods. The visual sensory apparatus in all these cases is epithelial and derived from ectoderm. In fact, *Pac6* expression in vertebrates is not limited to the eye, but extends to nasal placodes, parts of the brain and spinal cord (Li 1994; Amirthalingam 1995). In squid it is even expressed in the arms (Tamara 1997) and in sea urchins, in the tube feet (Czerny 1995). Minelli (2003) asks, "Is modern molecular genetic evidence more convincing than this?"

The very existence of master control genes has recently been called into question. Of the seven *Drosophila* genes that were originally considered "master control genes" for the eye, none have passed the criterion of producing homeosis, that is, gene expression when the gene is transferred from one site to another. Davidson (2001) expresses serious doubt about the concept of master control genes and calls them a "fantasy of earlier days." "The role of genes in morphogenesis is likely to always be indirect" (Minelli 2003). Shubin and Marshal (2000) go even further, believing that the specific role of the "so-called developmental genes" is rather mundane. "Most of them regulate the rate of cell proliferation or determine cell adhesion properties or act as transcription factor, binding to specific DNA segments. The scientific literature of the last decade is full of examples of genes whose temporally and spatially correct expression is critical for the production of complex, specific features, but at the time are known to encode for proteins whose mechanism is quite unspecific" (Schejter 1993).

The phenomenon of gene expression and a subsequent morphological event may be subject in interpretation to the logical fallacy called *post hoc ergo propter hoc*. It preceded it therefore it caused it.

Gene phenomena may be accounted for by the theory that the gene is produced as a result of mechanical forces acting on the cell, such as cleavage. Recent experiments have shown that distortive mechanical pressure on the cell induces gene expression. Emmanuel Farge, a physicist working at the Curie Institute in Paris, challenges the idea that genetic programs are necessary to twist, bend and bud into heads, limbs and internal organs. In *Current Biology* (2003) and reviewed in *Nature* (2003), he reports that gene expression can be modulated by the mechanical strains of morphogenetic movements. These morphogenic movements can be as simple as growth within a confined membrane. When Farge briefly flattened three-hour old fruit fly embryos, their surface switched on a gene called "Twist" which is normally expressed only on the ventral side. Within eight minutes of compression, Twist expression had spread to encompass the entire embryo, even in embryos in which the dorsal-ventral patterning system had been eliminated (Scott 2003).

Farge proposed that the compression of the embryo was inducing ß-catenin to move into the nucleus (Farge 2003). ß–catenin has a gene-activating function (Sanson 1996). Farge also showed that the mechanical deformation caused the translocation of Armadillo to the nucleus. Armadillo is a dual action protein, a transcription factor and a component of the ß cadherin adhesion complex at the plasma membrane. When the embryonic cells of the *Drosophila* were compressed during the first phase of germ band extension, which leads to anterior gut formation, Twist expression increased eightfold. "These cells in the embryonic *Drosophila* are normally subjected to pressure that is made by the extending germ band at the posterior pole and the pressure is transmitted to the anterior pole, compressing the cells against the invagination of the mesoderm and foregut" (Review Heinrichs 2003). "You can reprogram the embryo mechanically at the beginning of its life," concludes Farge. It is known that distortive mechanical pressure on the cell can induce cell division. For instance genes that strengthen the cells lining blood vessels can be activated by the stretching effect of high blood pressure. This mechanical stretching expresses the TIGR/MYOC and Oculomedin gene and subsequent protein expression (Tamm 1996; 1999). The TIGR mRNA/protein expression is essential to glaucoma pathogenesis (Nguyen 1998). This action is thought to occur by some interaction with the plasma membrane in order to sense the tension of the membrane. It is suggested that the mechanosensors are stretch-activated, thus sensing the mechanical stretch of the membrane (Hamill 1992; 2001; Sackin 1995). Another candidate for mechanosensors is integrins, transmembrane receptors that couple components of the extracellular matrix with the actin of the cytoskeleton, lying directly under and attached to the membrane (Wang 1993). The cytoskeleton may form a complex with membrane proteins such as ion channels, regulating their responsiveness to external forces (Watson 1991). Gene induction by mechanical forces is found also in cardiac cells. In cultured neonatal cardiac myocytes, mechanical stretch causes the activation of the synthesis of a number of proteins (for review see Hamill 2001).

Cell division is caused by deformation of the surface by compressive or tensile force acting on the cell, in the plane of the smallest circumference produced by the deformation (Plate 19). This is Sachs' rule of cell division. The common notion that cells divide at the instigation of a protein produced by a gene is inconsistent with the general way cells divide. Outside the special environment of embryo, cells divide haphazardly like the root of a plant, accommodating any constraints in its pressure to grow in the direction of least resistance. Embryology is the molding of the body by topological constraints on cell division.

This same induction of gene activity by mechanical pressure may also play a role in the induction of genes in segmentation. The observed patterns of segmentation may be accounted for as stress patterns imposed by the bending of the embryonic axis.

Something as simple as uneven growth rates producing an embryonic bend with bands of cells can be the generating force for gene induction (Huxley 1924; See LovTrup). Once the first set of genes is induced, their gene products will turn on the next set, the genes acting as switches much like the electrical switches in a house. However, studying the switches is not the route to understanding how the appliance works. The science of genetics by knock-out gene experiments is the same problem as determining which fuses operate which circuits in the house. This is done by opening fuses one by one, observing which room goes dark.

The importance of mechanical stress in morphogenesis is now widely accepted (Newman 2003). In tendons, the fibroblasts and the extracellular matrix are modulated by mechanical stress (Vogel 1989). During embryonic development, this mechanical action is particularly important in that the extracellular matrix for the formation of bone can only arise as a result of mechanical stress (Herring 1993). These same authors acknowledge that substantial stress is generated during embryonic growth.

While it is generally accepted that genes guide the assembly of proteins that constitute the body, autogenesis maintains that gene proteins are the result of mechanical forces on the cells as the tissue expands under constraints. Embryonic cell division is guided by the accommodation of the immediate forces of folding and stretching, and the genotypic details, which differentiate the individual or species, are the result of 30,000 on-off decisions of superficial Mendelian alternatives that vary the relative size of the parts.

In *Drosophila*, the egg experiences four stages of segmentation events triggered by four sets of genes: gap genes, pair-rule genes, segment polarity and Hox genes. The first segmentation caused by the gap genes is transitory. It is succeeded by consecutive segmentation patterns of the next two series of genes. These produce the final segmentation that consists of segments produced by the combination of anterior and posterior halves of consecutive segments produced by the previous gene group. Finally, Hox genes initiate appendage formation on the segments. Segments are distinguished by the narrow separating boundaries that produce different optical effects due to bending, refraction, pigmentation, and chemical and physical changes. Patterns observed can be rationalized as the serial appearance by self-organization of layers of tissue, each out of register with the next by the width of one segment, each producing gene expression in the order observed.

Mendel showed that superficial traits are inherited according to a choice presented by gene selection of one of two versions of the character, such as giant or dwarf, pigmented or not-pigmented, hard or soft, flexible or stiff (Mendel 1865). These are invariably simple on or off decisions between the more developed or the less developed version of the organ or organism, as though each is available in large or small size, in white or color. Given the vertebrate phenotypic body form, uniform throughout the entire sub-phylum, a whale or a hummingbird will result depending on an encoded program of on or off decisions. Large hind leg genes make for rabbits and kangaroos. The entire organismal form of the animal comes in two versions, male or female, one slightly tending more to neoteny than the other, all decided by a single set of genes on a chromosome.

Chapter 9 **Form Without Inheritance**

The extent to which the final outcome is presaged in the egg is an issue that has engaged those seeking to explain embryonic development since antiquity. According to the concept of preformation espoused by Charles Bonner, the new organism was already present in its final form in the egg, so that development simply entailed enlargement without any accompanying increase in complexity. During the latter half of the nineteenth century, preformation was revived in a subtler guise to explain the development of various marine invertebrates in which both patterns of cleavage and differentiation of the resulting blastomeres were essentially invariant. Here, cellular diversification was attributed to the localization within the cytoplasm of the egg of factors, or 'determinants', which dictated the fate of the cells that inherited them. The stereotypical pattern of cleavage exhibited by such organisms was referred to as mosaic. *(Gardner 2003)*

Stuart Kaufman's *The Origin of Order* (1993) argues that self-organization plays an important role in the emergence of life and may play as fundamental a role in shaping life's subsequent evolution, as does the Darwinian process of natural selection. "No systematic effort has been made to incorporate the concept of self-organization into evolutionary theory. The construction requirements which permit complex systems to adapt remain poorly understood." So far no one has detailed just how nature's beautiful patterns and symmetries are directly coded for in the genome (Miodownik 2003). A few biologists have even suggested that fundamental aspects of an organ's form persist despite natural selection, not because of it (Goodwin 1994; Ball 1999).

The transmission of structure and form to progeny is not dependent on the transmission of genetic material. There are many examples of natural phenomena that are reproducible without being encoded by any genetic program, says Allesandro Minelli in *The Development of Animal Form* (2003). There are default morphologies, much like an individual default fingerprint that is accurately reproduced over and over and dependent on only non-genetic cellular information (Fusco 2000). Cells grow and divide using the template of a preexisting cell. One centriole produces two using a preexisting template. There is a general property of cells that determines its axis, independent of any gene's instructions. For instance, yeast cells repeatedly polarize and will continually bud from a particular pole. The polarization is due to a membrane protein called Rax2 that takes a fixed memorized multigenerational position in the cell's cortex (Muller 1997; Chen 1999).

PLATE 32: *Crustacean Morphogenesis*

Cytoplasmic memory is well known in cells, in particular in that of ciliates, which repeat, generation after generation, perfectly patterned rows of cilia without any direct intervention from the genome (Tarter 1961). In the 1960s Tracy Sonneborn discovered that he could transplant bands of cilia on a *paramecium* by slicing off chunks of membrane with attached cilia and rotating the direction of the beating cilia. *Paramecium* divide by splitting in two and each of the daughter cells of Sonneborn's surgically altered *paramecium* had cilia that beat in the opposite direction. This surgically altered graft was carried into succeeding generations (Sonneborn 1965). No change in the DNA occurred, but each daughter cell now carried the pattern of the misplaced cilia. Another example cited by Minelli (2003) is the self-assembly of the multicellular alga *Hydrodyction.* This alga has a three-dimensional lattice structure made of many cells, which also serve as a non-genetic template for the three-dimensional pattern of the next generation (Bonner 1993). Minelli said there are many examples of organisms transmitting major morphological traits or developmental patterns by non-genetic means, but he questioned whether this was possible in metazoan animals due the size limits of the egg and the amount of information necessary for the patterning of the adult body plan.

Egg cells encapsulate themselves in a series of membranes that enclose the cytoplasm and an outer inelastic sphere, called by many names. This envelope is called zona pellucida in mammals, chorion in fish, vitelline envelope in amphibians, vitelline layer in echinoderms and perivitelline membrane in birds. These envelopes are semi-permeable, allowing water and small molecules to enter by osmosis, providing substance for growth and expansion of the inner membrane (Van der Hoef 1984). The inner cell membrane, a thin elastic shell is comprised of lipid molecules in hexagonal arrays (Lau 2002), expands uniformly with growth and cell division. The egg surface membrane, experiences a predicable sequence of deformations, including buckling, folding, creasing and crumpling and after many divisions finally enters its own interior, forming the embryonic shape (Plate 32).

With this inversion, the pattern becomes reversed, and the component segments subject to plastic deformation as they pass through the inversion aperture. In a few hours a cellular precursor of the complete vertebrate skeletal system is constructed. The transparent tissue of potential mesenchyme cells becomes visible in gradual stages only later during embryogenesis, like an exposed photographic plate in the developing tray, each bone revealing itself from the center outwards. According to the patterngenesis model and as confirmed by many recent experiments like the Keller sandwich explants, skeletal components do not form without gastrulation.

The encoded configuration of segments directs the development of the phenotypic body plan, the standard chassis upon which the genome may direct modifications. Given a vertebrate body plan, the genome can differentiate human from chimp, and from any other human, but it plays a role other than a creative role in directing morphogenesis. The DNA of a chimpanzee and a human has almost identical structural genes, almost 99 percent (King and Wilson 1975). Jeffrey Swartz (1999) expands on this:

The effects of the genetic regulation are [sic] more the shapes of the features, rather than the structural genes.... Clearly, even though they share virtually their entire suite of structural genes, humans and chimpanzees are members of different species.... If nothing else, the whole of the human fossil record—with what we now know to be numerous species of Australopithecus-like forms, as well as an increasingly recognized number of species of Homo—intervenes between them. This in turn suggests that all extinct hominid species would have shared most of their structural genes with humans and chimpanzees. But humans, chimpanzees, and the species of extinct hominids are remarkably, and in some cases almost unbelievably different from one another morphologically. (Swartz 1999)

There is confirming evidence for autogenesis in the correlation between hypothetical membrane patterns and body segmentation patterns in several bilateral animal phyla. This theoretical stress pattern generated on a spherical or ovoid shell by axial compression is in mimicry of the presumed physical circumstances of oögenesis or egg development. Homologues of any of the possible variations in the compression stress patterns can be found in corresponding configurations among species in all phyla. Conversely, the segmentation pattern of any species can be associated with a corresponding generating stress pattern. This theory leads to the inference that the mechanical stress pattern created by compressive forces during the formation of the egg determines the phyletic body form. This compressive force and subsequent buckling may be caused by growth within the constraints of the relatively inelastic vitelline envelope or it can be set in motion by a small imperfection in the sphere as in the buckling of a shell as detailed by Thompson (1982). The subsequent curling of the axis may be merely early variations of growth rates of the two sides of the elongating ellipsoid resulting from the inherent mechanical instability of the situation. We see this curl of an elongated ellipsoid in a pickle or a hot dog.

Additional evidence for vertebrate form resulting from compression patterns on the egg surface is the correlation of change in bilateral phyla evolutionary pattern with the changes that occur in the relative growth rate of the axes. Vertebrates, insects and crustaceans, all reveal in evolution: 1) a reduction in the number of segments, 2) a reduction of the number of segments bearing limbs and 3) condensation of groups of segments. This morphological direction, that is the reduction in number of segments, corresponds with the transition of the compression pattern produced by a more elongate to a less elongate spheroid. The contour of the egg is spherical. It elongates before establishing the body proportions. The persistent suppression of this tendency, that is, the retention of the earlier traits, neoteny, can account for the direction of the evolution of bilateral animals, the passage from an elongate to a spherical body.

This configuration of circumferential bands is the proposed origin of the segmented body plan common to all metazoan life. The segmented egg surface membrane, subject to simple, predictable topological maneuvers, produces the radial and bilateral phyletic body plans, known as the phenotype, the features shared by all individuals of the same phylum. The genome is now known to be responsible for the genotypic characters, those characters that differ among individuals within the phylum, such as relative size of the parts and differing chemistry.

Come and give me a nice
affectionate hug,
but gently, so the glass
won't break!
It is a curious property of things
that what is natural takes almost
endless space,
while what is not, requires
a container.

From Goethe's Faust Part II, where Wagner is addressed by his Homunculus

Chapter 10 **Patterning**

Whatever the precise mechanism of biological pattern formation might be...it takes place in the embryo not in the animal form. Even if the embryo does not exhibit clear patterns, they have to be present in some cryptic form. (Stewart 2001)

"Your world was shaped in the first twenty-four hours after conception. Where your head and feet would sprout, and which side would form your back and which your belly, were being defined in the minutes after sperm and egg united" (Pearson 2002). The development of an organism is the transformation from one set of patterns to another set of patterns. It is pattern formation that specifies this organization; it is pattern formation that specifies body segments, symmetry, polarity, and in vertebrate limb development even specifies where cartilage will form and how it will grow (Wolpert 2003). This pattern formation is positional information with cells acquiring positional identity and value in relationship to their boundaries. Positional identity means that the cells in the embryo seem to know where they are at all stages of development because of chemical signals. They can then use this information to follow different paths, depending on where they are.

Genes do not store the information on this position or map in space or time of a given cell during the development of an individual animal (Edelman 1988). Development is the emergence of organized structures from this pattern. When the fertilized egg develops, the resulting cells give rise to specific defined structures and appendages.

PLATE 33: *Vogt Fate Map*

Currently, there is evidence that this patterning originates very early within the egg (Gardner 1997, 2001; Zernicka-Goetz 2002). Patterning has always been of interest to developmental biologists, with many early researchers describing patterning in oöcytes. Many years ago, Lester Sharp in his *Introduction to Cytology* (1921) said, "...just after syngamy and before cleavage, the egg substances become arranged in 'patterns', which are characteristic of the various phyla." This patterning became known as "Fate Maps" and was the subject of research of many early embryologists. Plate 33 illustrates a graphic representative of one of Walter Vogt's fate maps (Vogt 1929). Vogt originally published his results with beautiful colored maps illustrating the developmental fate of specific cells in the urodele blastula (an amphibian). He applied harmless dyes into the surface of the early embryo and then traced the fate during embryonic development. Although Vogt's maps looked very convincing and are still seen today in many biology textbooks, the maps were inaccurate due to the diffusion of the dyes (LøvTrup 1966). Fate maps today are much more accurate and involve special dyes and techniques such as inserting microscopic hairs through growing embryonic tissue (Keller 1991).

Through the use of such maps during embryonic development, patterning can be traced as for back as the egg itself. The establishment of polarity is seen in the very early embryo (Zernicka-Goetz 2002; Edelman 1988). In the egg of the African frog, *Xenopus*, the organization is both bilateral and radial. A single axis passes from the animal pole to the vegetal pole. The development of all low-order animals follows an organized, polarized directional course from fertilization through fetal development (Scott 2000). In most species this polarity of the embryo takes its roots from the spatial patterning of the egg.

For a long time, early patterning in mammals was viewed as the exception. However, Jean Smith noted that the mouse blastocyst was slightly distorted and had

PLATE 34: *Shell Instability: "Induce instability in a spherical shell and allow it to buckle spontaneously into a circular dent, collapsing in on itself. This scenario comes free of charge to frog biology. Frog genetics may trigger the instability and if necessary, it can evolve to modify the precise buckled form, but it doesn't need to tell the blastula how to buckle. Similarly the genes for an elephant do not have to tell it to obey the laws of gravity if it falls off a cliff. This means that you cannot understand gastrulation in purely genetic terms. You must build in the mathematics of buckling, too, or more precisely, you must build in whatever analog of buckling applies to the material of a blastula" (Stewart 1999). Photo printed with permission from John Wiley from* Catastrophes in Science and Engineering *by J.M.T. Thompson, 1982.*

PLATE 34

PLATE 35: *Membrane Stripe Out (Baumgart 2003)*

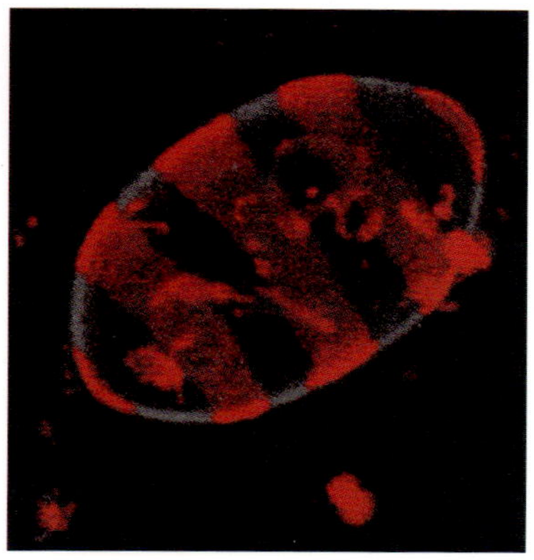

a recognizable axis that matched the fetus (Smith 1980). Magdalena Zernicka-Goetz also working with mouse embryos, said, "there is a memory of the first cleavage in our life and the first division of the egg influences the fate of each cell and ultimately all the tissues of the body" (Piotrowska 2001). Zernicka-Goetz and other researchers like Gardner and Davies (2003) also concluded that the axes of the mouse embryo could be traced back as far as the egg itself. They wrote that the early mouse zygote has a bilateral symmetry before regaining an essentially spherical shape prior to first cleavage, and that the bilateral symmetry is a manifestation of the intrinsic organization of the egg. The polarization begins in the unovulated ova and is reinforced at fertilization. After fertilization the axis and polarization endure to the blastocyst state and later define the axes during gastrulation and fetal development (Scott 2000). Gardner demonstrated there is spatial distribution of cells in the visceral endoderm that can be traced back to the polarity present at the blastocyst stage (Tam 2001; Gardner 1997). This was originally proposed by Huber (1915).

There are at least two spatial cues that operate in the mouse egg to shape polarity of the blastocyst. One of these is at the animal pole, which is defined by the site of the female meiosis, and another is associated with the position of the sperm entry. This genesis of polarity in the early embryo is crucial to development because it determines the body plan of the whole animal. The sperm penetrates the mouse oocyte preferentially at regions other than the animal pole (Evans 2000). Phytohemaglutinin-coated beads used to mark the egg surface where the sperm entered tend to retain their position relative to other surface markers from egg to blastocyst stages (Piotrowska 2001). Similarly oil drops placed in the zona pellucida (vitelline membrane) traced the axis of the growing mass of cells repeatedly through division (Gardner 2001). Johnson, in a wonderful review of this work, suggests there is patterning in the cortex. Johnson questioned the massive membrane addition and expansion during cleavage, with most of it going deep into the embryogen during cellularization (Johnson 2001). *The beads did not relocate to these interiorized membranes*, which suggest that there is a surface matrix within the membrane or cortex encoding developmentally significant spatial or temporal information. This cortical encoding has also been suggested by a number of other scientists including, Pratt (1989) and Louvet (1996) in the mouse, Roberts (1992) in *Xenopus*, and Lecuit (2000) in *Drosophila*. Johnson (2001) interprets this response to indicate the presence of some "solid islands or even continents linked through some underlying cytoskeletal structure." Although Johnson does not go so far as to say this, these solid islands within the oöcyte membrane could very definitely be the membrane patterns predicted by the theory of patterngenesis. He further went on to say that whatever their structure, some surface features of the egg appear to be preserved throughout these developmental stages. Johnson questioned if this patterning of islands could be created in a sea of fluid membrane lipids, a patterning encoded within the membrane or the cortex.

The famous cell mimics made by Russian biologist Oparin in the 1920s, consisting of artificially produced lipid bilayer spheres, have been observed by Baumgart et al (2003) to generate surface patterns of self-organized circumferential zones of the type of the hypothetical vertebrate pattern. Using high-resolution fluorescence imaging dyes, Baumgart and his group were able to observe long-range domain ordering in the form of locally parallel stripes and hexagonal arrays of curvature-dependent circular domains. The authors referred to these circumferential zones on the membranes as "stripe-out" (Plate 35). Their prepared bilayer lipid membranes showed definite arrangements of both hexagonal and inverted hexagonal patterns. Depending on membrane composition, they also observed multiple stripe domains arranged laterally in undulating patterns of locally parallel stripes. "The stripe-out (on the membrane) that we describe here resembles the rippling instability of collapsing bubbles.... The lipid bilayer membranes—ubiquitous in biological systems and closely associated with cell function—exhibit rich shape-transition behavior" Lau (2002) found this same lattice-like array of particles in a repeating pattern on the *Xenopus* oöcyte plasma membrane (See review on membrane patterning in Muller 1997).

As noted in an earlier chapter, an artificial phospholipid bilayer forms a torus. This was recently confirmed by Jülicher, Seifert and Lipowsky in their work with shape transformations of toroidal vesicles (Jülicher 1993).

These lattice-like arrays of particles or hexagons have been documented on membrane surfaces created by the packing of the phospholipid. Hexagons are the natural shape assumed when molecules are packed together (Ball 1999) (Plate 48). Phospholipids move easily within the cell membrane. Just jiggling a solution by sound waves will create the self-assembly of phospholipid membranes.

Mechanism for the morphogenetic event of pattern formation in tissue development is mechanical instability, or buckling. The first example of mechanical instability encountered in Mechanics is the column in which only small deflections are considered. If the column is simply supported at both ends, it will deflect from its straight shape into the shape of a half-sine wave when the compressive axial load acting on the column reaches a critical value called the critical load. There are other shapes that it will take as the compressive load is further increased. (Cowin 2000)

Mathematician and biologist Ian Stewart writes that circumferential bands forming patterns could result from buckling under compression failure (Stewart 1998) (Plate 34). According to Stewart, exactly the same buckling happens if uniform forces directed toward its center compress any kind of spherical shell such as a Ping-Pong ball.

Induce instability in a spherical shell and allow it to buckle spontaneously into a circular dent, collapsing in on itself. This scenario comes free of charge to frog biology. Frog genetics may trigger the instability and if necessary, it can evolve to modify the precise buckled form, but it doesn't need to tell the blastula how to buckle. Similarly the genes for an elephant do not have to tell it to obey the laws of gravity if it falls off a cliff. This means that you cannot understand gastrulation in purely genetic terms. You must build in the mathematics of buckling, too, or more precisely, you must build in whatever analog of buckling applies to the material of a blastula.

Micro-geometry would predict that banded segments on the surface of the egg are each extended hexagons derived from the regular hexagonal plates that pave the patterned primordial torus. With expansion of the surface, the hexagonal elements rearrange themselves, as surface tension seeks a minimal energy solution to the problem of the paving of the sphere with hexagons. This geometric impossibility leads to surface discontinuity, which can instigate instability. The vertebrate body results from the ingression of the surface into the interior. The surface ends up internally as bones, during the plastic deformations caused by passage through the blastopore. What is exoskeleton in arthropods becomes endoskeleton in vertebrates.

The congruency between the bilateral segmentation patterns of animals with the projections of the theoretical compression patterns on the spherical oöcyte surface may be used to establish the causal relationship between the two phenomena, that is, phyletic form is determined by the self-organized segmentation patterns on the surface of the egg (Plate 36).

PLATE 36: *Segmentation Congruency: The congruency between the bilateral segmentation patterns of animals with the projections of the theoretical compression patterns on the spherical oöcyte surface may be used to establish the causal relationship between the two phenomena, that is, phyletic form is determined by the self-organized segmentation patterns on the surface of the egg.*

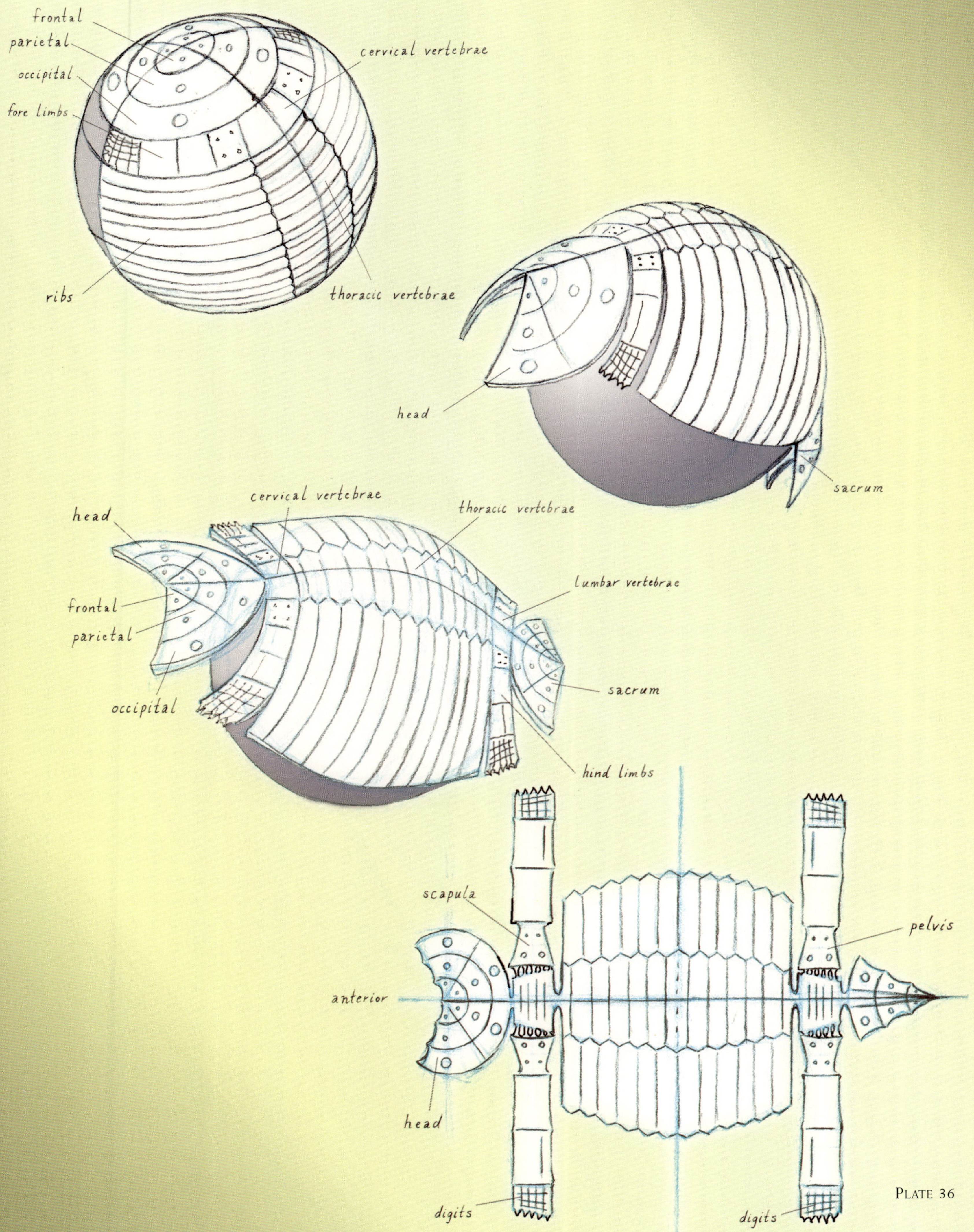
frontal
parietal
occipital
fore limbs
cervical vertebrae
ribs
thoracic vertebrae
head
sacrum
head
cervical vertebrae
thoracic vertebrae
lumbar vertebrae
frontal
parietal
occipital
sacrum
hind limbs
scapula
pelvis
anterior
head
digits
digits

PLATE 36

Chapter 11 Segmentation

It seems that eggs are patterned and compartmentalized not by a single, global mechanism, but by a sequence of rather cruder processes that achieve their goal only by virtue of their multiplicity. (Ball 1999)

From a developmental viewpoint, segments are something different than the segments one sees in millipedes, centipedes, armadillos or pill bugs (Plates 37, 38). Minelli (2003) suggest that segmentation should follow the definition of Alberch and Kollar (1988) in defining "segmentation as the subdivision of an embryonic field into sharply defined populations of cells linearly arranged along the anterior-posterior axis." Segmentation is almost a generic property of metazoan organization. "Segmentation repetition of structure becomes a built-in developmental potential of all animals" (Newman 2000). These segmentation patterns differ fundamentally in different groups. In some groups, they appear synchronously, and in other groups, they appear in an anterior-posterior sequence. In insects, the segmentation is ectodermal and in vertebrates, it is mainly mesodermal (Wilmer 2003).

PLATE 37: *Pill Bug*

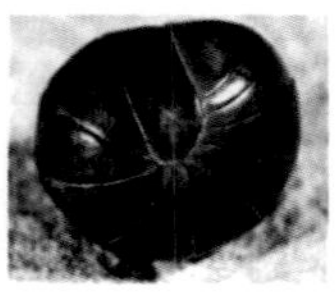

PLATE 38: *The shell of the armadillo is in the form of the predicted articulated sphere of the egg membrane, compressed dorsally. The armadillo shell is an enlargement of the mature phenotypic vertebrate pattern. It is the literal enlargement of the articulated sphere of its own egg.*

The well-known evolutionary trend in bilateral morphology is the passage from what are designated the homomonous (like segments) to the heteromonous (differentiated segments). The Cambrian fauna and other initial forms were characterized by uniform, serial, metameric segmentation. During later periods, segments became fewer in number and differentiated in form, as seen today in most animal forms. For example, the vermiform, multipodal caterpillar turns into a trimeric hexapod. The change consists of the condensation of groups of segments that become limb-bearing segments, or limb girdles.

In a paper entitled "Is Segmentation Generic?" Newman (1993) defends a hypothesis that the "capacity of undergoing segmentation may be 'generic' to tissues, that is, an outcome of their most general physical and chemical properties, without regard for any differences in the molecular makeup." He suggested that segmentation is almost a generic property of metazoan organization, an inevitable consequence of cellular molecules' auto-regulatory tendencies. It is Newman's view that segmentation organization is frequently based on the inability of otherwise similar tissues to exchange cells because of some type of interface, presumably physical.

At the earliest observable stages the egg appears elongate, segmented and its axis curled, with a band of transverse grooves across the midline of the inside curve

PLATE 38

called the germ band. D'Arcy Thompson (1917) summarized a number of phenomena and sequences of events to account for segmentation including the effect of harmonic resonance, the segmenting effect on an elastic body of mechanical energy provided by pulsation or vibration. Polymerization of the membrane protein along the fault lines can generate lines of visible force. The oöcyte grows during oögenesis within an unyielding spherical envelope that can cause the presumed circumferential buckling and axial segmentation along the force lines. With increased axial growth, the egg is forced to curl within the constraining sphere. Increased axial growth within the constraining envelope could cause the segments to crumple and compress within a deep furrow, and fuse as flat tubes that cross the midline, as the edges of the furrow close over them, pressing them together. With increase of girth, the axis straightens, the midline splits and separates laterally, and the formed limbs emerge, stand forth from the surface, achieving the posture of the arthropod and amphibian limb configuration, an arched tunnel that elevates a vermiform body above the ground, and can manipulate its attitude in all planes.

The adult insect body is subdivided into head, thorax and abdomen. Other arthropods, including spiders, scorpions and crustaceans, combine head and thorax. Vertebrates combine abdomen and thorax. Limbs appear from compressed thoracic segments. Insects only produce membranous wings by a presumed extra band of segments at an early stage that retreats, folded, only later to emerge fully formed. A meridian in-folding completes the internal body of the arthropods.

Embryologists, especially following the observations in the 1880s of mitosis (Plate 18), intensively studied the causes of segmentation of the egg. The main cause of segmentation is presumed to be mechanical failure under compressive or tensile force. The patterns established by these failures, such as buckling and folding, may well be preceded by stress patterns established in the sheet due to other mechanical causes. This subject was summarized by D'Arcy Thompson (1917), who presented the theories of the causes of cleavage and segmentation in the early egg by various authors attributing it to electrical, hydrodynamic and chemical causes. He cited the work of Ida Hyde (1904), who discovered and measured the oscillating micro-voltage potential in the dividing cell. Thompson also discussed the creation of a standing wave by the mechanical and chemical effects of pulsations, or vibrations in the gelatinous mass of the cytoplasm producing compartmentalization. These experiments in mechanical modeling were continued by later workers, including Lewis Wolpert (1969) and Philip Ball (1999).

During oögenesis, the egg interior is in turbulent motion because of cytoplasm streaming. This movement of the cytoplasm and the accompanied tension and movement of the microtubules, attached both to the cell cortex membrane complex and the centrosome of the polarized egg, produce pulsations and vibrations that distort the membrane surface. As surface is distorted under compression, the accordion folds and crimps are alternately opened and closed by the pulsation of the axis. The rapid folding and unfolding of the surface, each time more advanced than the last, could provide the energy to polymerize the cortical protein along the lines of force, producing visible bands. These are the guiding stress failure patterns for a compression failure to follow, which can produce internal circumferential hinges. The correlation between segmentation pattern of the phyletic Bauplan and the theoretical axial compressive stress pattern of the spheroidal cell from which it was generated suggests that the self-generated patterns on the egg surface are the origin of the phyletic Bauplan.

Segmentation begins superficially and grows deeper in ontogeny and phylogeny. In vertebrates, segmented skin patterns are not related to the patterns of the ribs beneath, which in turn have nothing to do with the configuration of the internal organs. The segmentation of the insect abdomen consists only of a hinged shell. Segmentation may be deeper, to the degree that in some lower phyla segmentation is total, each segment bearing a separate set of internal organs like so many train cars. The segmentation patterns within groups are consistent. Insects have twenty segments, six for the head, three for the thorax, and eleven for the abdomen. Millipedes have hundreds; lobsters and crabs have a five-segmented cephalo-thorax and a tail of six segments. Humans have a tri-segmented skull and twelve ribs. Birds and crustaceans have fantails, which can easily be imagined as the result of the radial symmetry of the south pole of the egg, which produces the telson and fantail, as well as the sacrum in vertebrates. The arthropods include innumerable species exemplary of the segmental arrays and axial contours predictable by the algorithm of the stress pattern upon the surface of an ovoid. The subdivision of the serial segments into body sections follows the universal plan of head, thorax and abdomen.

This model of spontaneous generation can predict the morphogenesis of most bilateral phyla, although in most marine invertebrate phyla, embryogenesis includes a temporary larval state. Larval forms are visible as millimeter size, complex, self-sufficient animals in astounding variety, in hundreds of different types (Hall 1999). Larval forms are temporary stops in morphogenesis and are irrelevant to the general morphogenesis of the adult.

Chapter 12 Spherical Origami

Complex systems can arise from simple causes. Given the simplest instructions a computer can generate patterns resembling trees and leaves. Conceivably then, the general, geometric format common to animal life may spring from one simple algorithm that guides tissue growth, forming the regular, symmetrical patterns that occur over and over in diverse parts of nature.

The homunculoid algorithm provides a solution to the mathematical/biological problem of long standing: how to transform a spherical cell into the shape of the animal body (Plate 39). That such a formula should exist is suggested by the obvious regularity and ubiquity of certain shapes in nature, especially apparent in simpler organisms such as diatoms, mollusk shells and flowers. Fleas, lobsters, elephants and man have the same fundamental body plan, that is, a segmented bilateral semi-cylinder, with head, torso and tail, paired, jointed limbs, an anterior face with a pair of dorsal eyes and an alimentary canal with ventral anterior and posterior openings (Plate 40).

The algorithm consists of a sequence of simple deformations, which convert the surface of a spheroidal torus into the body plan of a plant or animal. The enclosing spherical membrane of the fertilized egg is seen to be empatterned much like a terrestrial globe with parallels, polar caps and intermediate zones. The figure is composed of quadrilateral surface elements. Each is a membrane capable of becoming a particular morphological feature within the final form.

Subdivision of the membrane of the fertilized egg results in the blastula, a thousand cell mosaic version of the egg itself. The blastula is a replica of the egg, the surface subdivided into a thousand separate compartments. It is known that the cytoplasm of dividing cells remains connected by thin tubes called plasmodesma. These dividing cells acting in concert as a tissue form a plasmodium or syncytium, a single membrane that encloses a multiplicity of nuclei. The plasmodium is the general form of the cell in the simpler phyla. "Plasmodium Theory" sees the body as a single mass of protoplasm consisting of more or less interconnected cells. This view of a mass of cells acting as one unit was demonstrated in Piotrowska's work (2001) with the phytohemaglutinin-coated beads, marking the egg surface and retaining their position relative to other surface markers from egg to blastocyst stages. This retention of site persisted despite a thousand fold increases in membrane surface. Lester Sharp characterized groups of cells acting as a "plasmodium" in his 1921 text *Introduction to Cytology*:

PLATE 39: *Somite Formation. Distal segments reverse curvature as they pass through the outlying part of the vortex, folding into bales, which line up in pairs alongside of the forming vertebral column. These unwrap and elongate, forming the tissue for the future ribcage.*

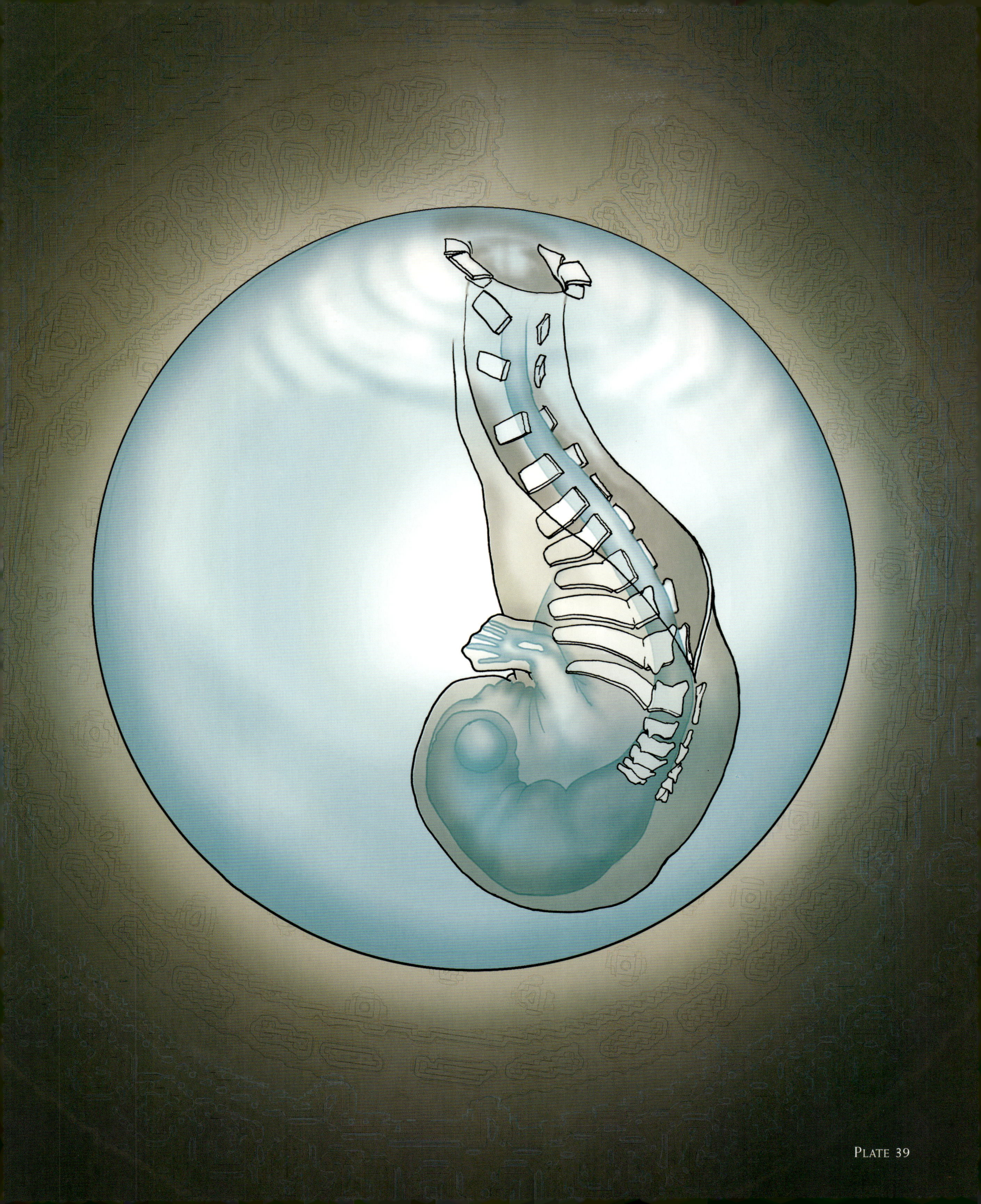

. . . the multicellular plant or animal is not a colony or republic of elementary cell individuals, but rather a more or less continuous mass of protoplasm (plasmodium) which has become incompletely subdivided into subordinate centers of action, the cells, during the course of ontogenesis; the cells are the result rather than the cause of development and differentiation. In the words of deBary, "die Pflanze bildet Zellen, nicht die Zelle bildet Pflanzen." (The plant forms cells, not the cell forms plants). *The phylogenetic corollary of this theory is that multicellular organisms have evolved not by an aggregation of many individuals, but rather by the growth, differentiation and septation, i.e., as by just such a process as is observable in the ontogeny.*

Subdivision of the egg into separate cells is an exception rather than the rule of embryonic subdivision. Complete cleavage occurs in mammals, mollusks and echinoderms. Fish, reptiles and arthropods cleave superficially, with the nucleus only dividing and no actual cell division until a later time (Gilbert 2003). The insect egg is one of the best examples of no early cell division (Trougakos 2002). The nuclei proliferate inside the insect egg, and suddenly at a point just before they migrate to the surface, boundaries form around each of them, effectively subdividing the surface into separate cells (Hilker 2002; Lawrence 1992; Snodgrass 1993) (Plate 41). Subdivision, by superficial cleavage or by binary fission, establishes the system of cell proliferation, which results in the enlargement of the embryonic form and generates variations by different rates of growth of the parts.

Cellular Holography

This process of subdivision by binary fission confers a holographic character to the resulting mosaic form of the embryo. In the first cleavage of the fertilized egg cell, the original cell becomes two daughter cells, the next division producing another two near-replicas of the original cell. As cleavage continues, each of the blastomeres, the cells of the blastula, embodies a replica of the form of the whole. The result is a hologram of the fertilized egg cell. Morphology results from the geometry of the cleavage ancestry, which confers a unique geographic position, like a piece in a spherical jigsaw puzzle.

In many phyla any one of the first eight cells of the hologram can produce an adult. Regeneration is a fundamental property of the cell membrane (Gilbert 2003). Each of the blastomeres, the newly formed cells resulting from the subdivision of the egg surface, senses through molecular signals or direct connection its unique position among its neighbors. If disarranged during cleavage, through their mobility they find their way to their former neighbors by chemotaxis, the pursuit of chemical trails in the ambient.

The holographic character of the cells of the multi-cellular animal may account for the phenomenon of regeneration, common mainly to simple animals. Cells bordering the excised sections, each a hologram of the whole, reproduce what is their normal neighbor. Each piece of a chopped-up hydra or starfish regenerates a whole animal. Both halves of a bisected flatworm reproduce the missing half. Some amphibians can re-grow limbs and the gecko, its tail. We can regenerate small wounds, our liver and even a fingertip. Cells comprising a larger portion of the original egg membrane surface are more capable of regenerating lost tissue. The secret of the newt's ability to regenerate body parts lies in ordering specialized cells at the wound site to revert to embryonic state and using these embryonic cells to build a new limb from scratch. These embryonic cells actually form a sphere of precursor cells called a blastema, which is able to grow, multiply and re-specialize to make any of the tissues needed (Merchant 2000).

The regenerative ability of biological forms has been compared to the well-known reparative properties of inorganic crystals. A known phenomenon of crystallography is that crystal with a void or a broken point will regenerate the defect before continuing general growth (Haraway 1976).

In many insects, neither wings nor other appendages grow continuously. Instead, limbs, mouthparts, wings and antennae appear from flat imaginal discs that stud the surface of the egg. These suddenly telescope forth producing the completed limb. The mechanical explanation of imaginal discs is that they are the result of the compression at an early stage of the formed appendage by the mechanical constraint of an unyielding envelope.

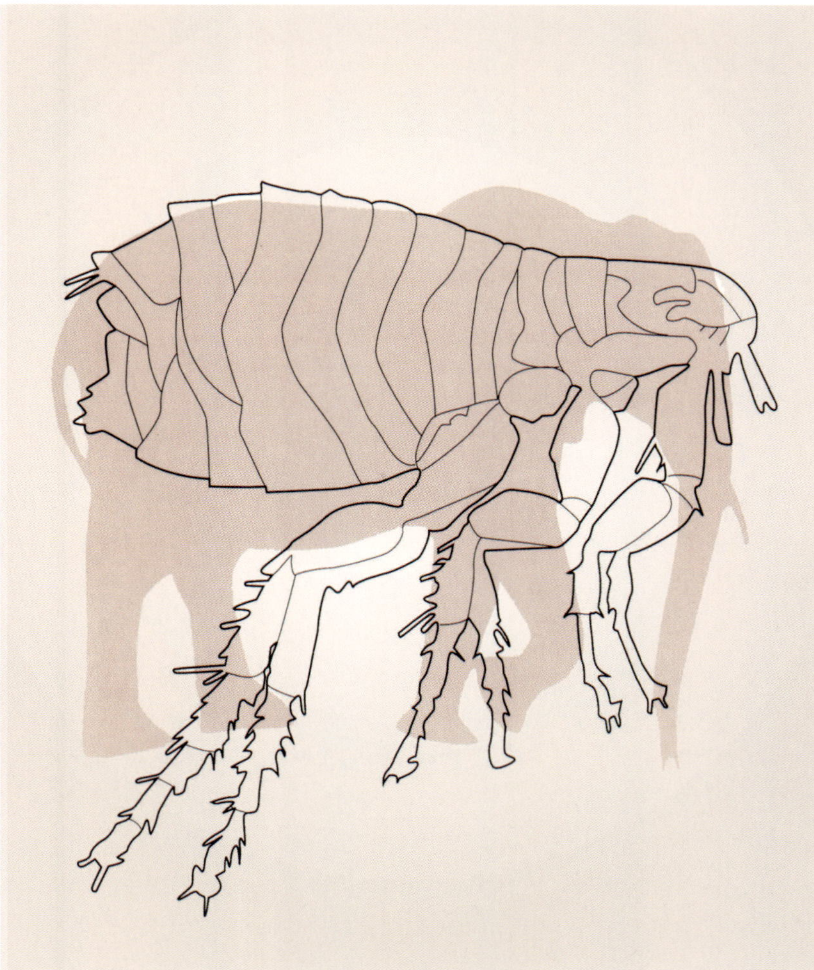

PLATE 40: *Elephant and a Flea from Ian Stewart, "What Shape is a SnowFlake?" (2001) Copyright Ivy Press*

PLATE 41: *The insect egg does not undergo cleavage. The nuclei proliferate inside the insect egg. They migrate to the surface. Boundaries form around each of them, effectively subdividing the surface into separate cells.*

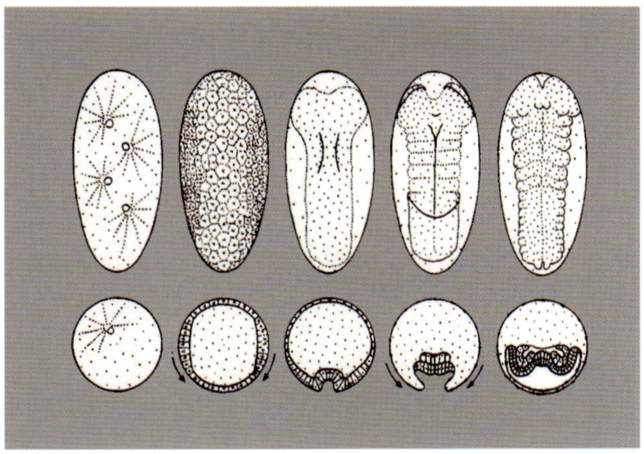

Chapter 13 Mechanical Engineering

The laws of physics...imply that a spherical shell is a natural stable configuration for a collection of small particles, weakly attracted to each other. They also imply that give or take a few technicalities, that whenever such a spherical surface buckles, it does so by growing a circular indentation and folding inside.... Maybe the individual cells of a blastula influence the buckling. Maybe genetic instructions tell those cells to start migrating...nevertheless, a significant piece of what then happens comes from physics, not from genetics.... The organic world is just as mathematical as the inorganic world. The mathematical basis of living things, however, is much more flexible and more deeply hidden. It is not just a matter of writing down some simple elegant Newtonian laws of life. (Ian Stewart 1998)

How do great oaks from little acorns grow? The history of biology is pervaded with unsuccessful attempts to rationalize the science with a universally accepted, fundamental theory to explain how the complex organism arises from a virtually invisible speck of matter. Biology is set apart from physics and chemistry by this failure to explain the mechanism that drives morphogenesis and relegates its study to the merely descriptive.

Many different approaches to solve the sphinx-like riddle of morphogenesis have been undertaken in the twentieth century. Some were by scientists in fields other than biology, notably physicists and mathematicians, who do not have a vested, professional interest in the theory of genetics. "Organization dynamics of morphogenesis define a limited number of points in space; the possible range of biological form is restricted in a fundamental way" (Stewart 1998). Goodwin (1994) also acknowledges the importance of mathematics in developing form and encourages its application to the study of embryogenesis.

Making Sense of Life by Evelyn Fox Keller (2002) recapitulates the leading twentieth-century attempts by Stephan Leduc, D'Arcy Thompson, Alan Turing and Nicolas Rashevsky to provide many possible non-genetic explanations for development based on physical, mathematical or otherwise non-genetic phenomena. These theories have never been disproved or superceded. They are simply ignored because most biologists are now preoccupied with the genetic paradigm.

In engineering, mechanical objects such as architectural structural elements, housings, containment vessels and others are tested for material strength and the failure patterns recorded, analyzed and rendered mathematically predictable. Engineers refer to buckling, folding, creasing and crimping. These phenomena are employed in industrial manufacture by various techniques referred to as metal forming, which take advantage of the temporary fluid state assumed by materials during mechanical failure. To perform a rigorous mechanical analysis of embryonic development, it is necessary to replace qualitative observations with standardized engineering measurements. Mechanical engineering techniques have been used a number of times to study biological processes at the tissue and cellular level (Moore 1995). Keller's group has frequently used biomechanical techniques to assay changes such as the elasticity of the blastula envelope, particularly the stiffness of the converging blastopore lip (Davidson 1995, 1999; Moore 1995). However, to our knowledge, except for the speculation by Ian Stewart as detailed above, no biologists have examined the blastula as a thin-walled shell susceptible to buckling phenomena as described in structural mechanics (Thompson 1982; Hunt 1983, 2003) (Plate 35).

Internal circumferential invagination bands produce a "living hinge," a working flexible joint caused by the tensile failure of the band. A plastic element will elongate under tension by "necking," the formation of a thinner, extended form composed of molecules that have oriented axially. This oriented form has increased tensile strength and is the basis of wire drawing and the extrusion of plastics. The living arthropod exoskeleton is a shell composed of the protein alpha actin. Molecular orientation produces the beta form of actin, which is strong and flexible. This shaped form may be observed in the segments of a crustacean where the segmented sphere gains remarkable freedom of motion (Plate 42).

As the axis curls, it compresses the segments along the concave side, producing a series of collapse failures. Segments fold axially along the ventral midline toward either the anterior or the posterior. They lie flat and fuse as segmented tubes as they are forced into a deepening furrow along the midline. As the edges of the midline approach tangency, the compression encapsulates the folded segments, which confer the contour of closely packed, elastic, plastic, adherent surfaces. Growth increases the circumference, causing the flattened tubes to stretch to tensile failure at the midline, separating in pairs of hollow, pointed extensions from the spherical surface. The resulting bending movement also results in the crumpling of the appendages into several segments that remain joined by joints of deformed sheet (Plate 42). Compression, tension and torsion of thin-walled plastic tubes will mimic these same joints within a deformed sheet. Separation causes the axis of the extended appendages to lift from the surface and distend at an angle from the surface.

The action of the Hox genes in controlling the formation of appendages to the segments can be accounted for this way: the genes that create the stages of segmentation are expressed simultaneously the length of the egg axis. The Hox genes are expressed sequentially from anterior to posterior instigating the formation of mouthparts and legs anterior to posterior. A mode of collapse failure of a thin-walled cylinder is the axially sequential collapse inward of arched grooves formed on the inside curve of the bent figure, the failure pattern directed by the previously established segmentation pattern formed by the action of the previous gene groups. The sequence of gene expression observed is the result of the mechanical pressure on the cells along the ventral midline, and is an archival record of the event to be repeated in the next generation by the expression of genes that confirm and preserve the timing of the mechanical events essential for mechanical integrity of the surface.

The competition for space by cells growing in a confined area is a source of gene-inducing mechanical force. Although a structure by the elasticity of its envelope may vary the proportions of it parts, the volume of contents to fill them is constant. In tetrapod vertebrates there is usually a consistent relationship between limb size reduction and increase in vertebrae count, likewise between head and torso size. Large flowers produce small fruits. Winged insects have small legs. (Carroll 1995; 1999). In scarab beetles, the production of horns reduces the size of neighboring body parts including antennae, eyes or wings (Emlen 2001).

Somewhere out there in the as-yet-unrealized space of future mathematical concepts, there are some extremely important ones about the nature of emergent form. (Stewart 2001)

PLATE 42: *As the arthropod axis curls, it compresses the segments along the concave side, producing a series of collapse failures. Segments fold axially along the ventral midline toward either the anterior or the posterior. They lie flat and fuse as segmented tubes as they are forced into a deepening furrow along the midline. As the edges of the midline approach tangency, the compression encapsulates the folded segments, which confers the contour of closely packed, elastic, plastic, adherent surfaces. Growth increases the circumference, causing the flattened tubes to stretch to tensile failure at the midline, separating in pairs of hollow, pointed extensions from the spherical surface. The resulting bending movement also results in the crumpling of the appendages into several segments that remain joined by joints of deformed sheet.*

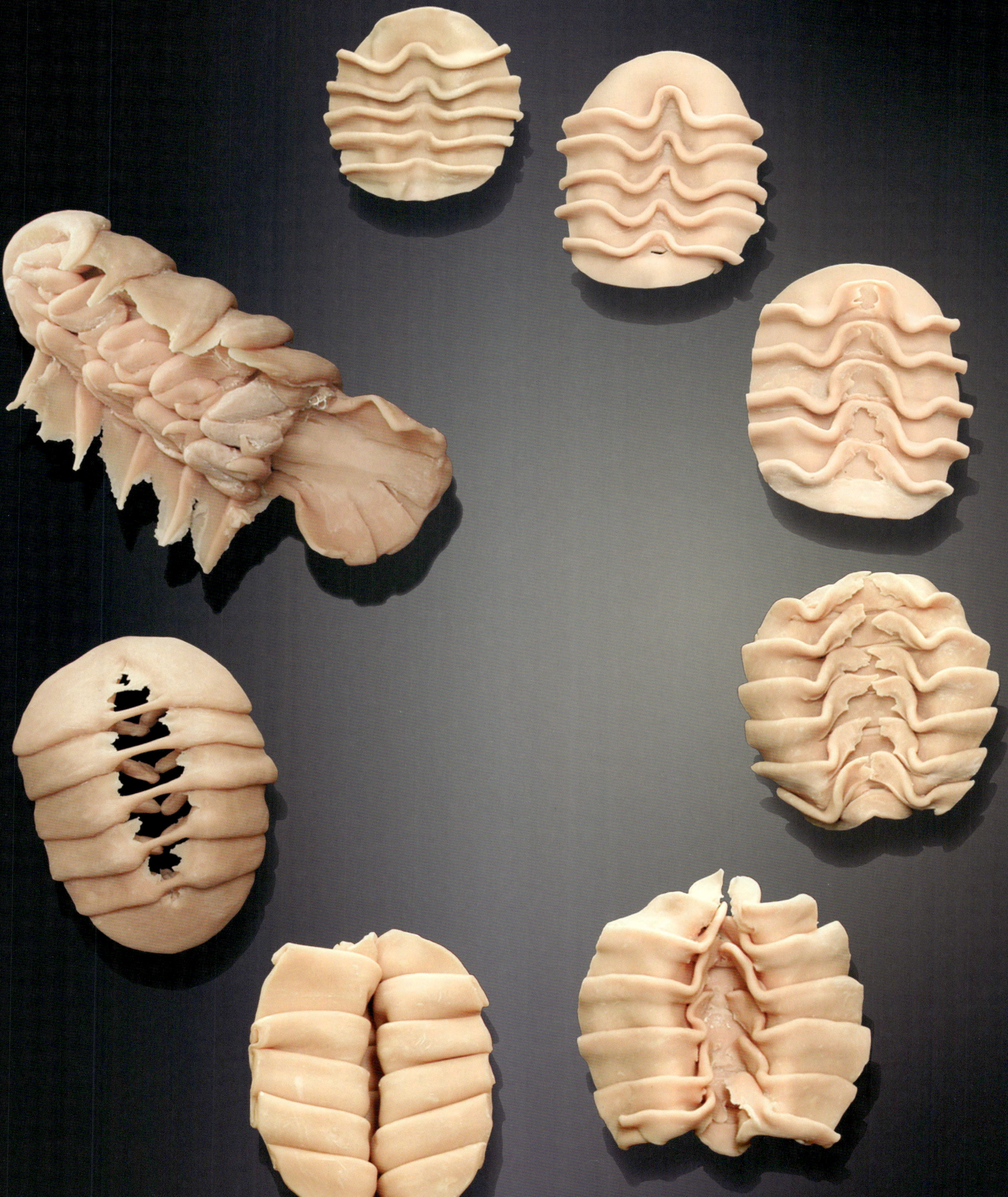

PLATE 42

Chapter 14 Eversion

In the embryos of most multicellular animals, sheets of cells invaginate during gastrulation, neurulation and organ formation. Gastrulation is the central process in early animal development, and occurs during the blastula stage—when the embryo consists simply of a hollow ball of cells. It involves complex movements that carry those cells, whose descendents will form the future internal organs from their superficial positions on the blastula to their definitive positions inside the embryo. Similarly, neurulation in vertebrates involves a complicated curling of a cellular sheet to form the neural tube, which, in turn develops into the central nervous system. (Nishi 2003)

The process of gastrulation or eversion is a universal biological phenomenon. In an article called "How to Turn Inside Out" published in *Nature*, July 2003, Schmidt and Sumper cite the recent discovery of a mechanism that causes the embryo of *Volvox*, a simple spherical marine organism, to turn inside out as part of its formation (Plate 43). They speculate that the discovery might have implications for similar events in animal development. This idea led Nishii and colleagues to reexamine and dissect the process of inversion in Volvox. Their concept helps explain the biomechanics and driving forces behind the curling of a cellular sheet.

Something remarkable happens to embryos of the multicellular green alga Volvox carteri. *They turn completely inside out to establish the adult body plan. This inversion process closely resembles the initial stages of the more complex gastrulation that occurs in animal embryos. So Volvox could arguably be considered a simple model for analyzing principles involved in such changes of shape* (Nishi 2003).

Inversion or eversion, is a common lifestyle among cells and simple organisms that evert themselves almost instantaneously. It is common for cells to pulsate in and out continuously. Eversion is the general behavior pattern of the cells of the radial animals. Marine metazoans feed this way, as do many of the higher phyla, in the mechanism behind biting, swallowing and regurgitating. The sting of the jellyfish is the harpoon of the everting nematocyst.

The propensity of the cell membrane to turn itself inside out results in the topological inversions that define the body plan. This derives from the intrinsic potential for an elastic bi-layer surface to reverse its curvature, the interior surface necessarily under compression and the outer, under tension. A spherical sheet, or ball, is in unstable equilibrium; its inside wanting to be out, and its outside wanting in. This work was originally developed by Stephen Smale, the mathematician who was said to have broke the dimension barrier in 1957, when he proved that it is possible to evert a sphere (2000). Given a discontinuity, the elastic energy stored like a spring in the inner layer will cause a split in the surface followed by the separate inversion of the parts. A rubber ball when punctured may begin an equatorial slit, while each hemisphere turns inside out. Many kids have seen this demonstrated by throwing a dead rubber ball into a campfire, and watching how, like a thing alive, the equatorial slit widens as the hemispheres turn inside out, remaining joined by a hinge. With impish perversity nature made protoplasm, which forms a bi-layer crust, thus beginning an eternal geometric strife during which neither side is satisfied where it is.

PLATE 43: *Schmidt and Sumper (2003) cite the recent discovery of a mechanism that causes the embryo of* Volvox, *a simple spherical marine organism, to turn inside out as part of its formation. They speculate that the discovery might have implications for similar events in animal development.*

Chapter 15 Embryography

Stripes in fur, stripes in flesh, stripes in sand, stripes in water. Is the similarity between them no more than a visual pun? Or is there a hidden unity? (Stewart 2001)

Nature keeps meticulous records. One can look up life forms of any geological age by referring to its strata. Glaciers, ocean sediment cores and tree rings record the climate and weather. Civilizations or their remains are preserved in stacks. A tree retains the archive of its life as coaxial cylinders. The outermost layer of the bark is a mosaic enlargement of pattern of the trunk in the first year. Similarly, the vertebrates retain a graphic record of embryonic stages in the patterns in the scales, feathers or fur of their hides.

Patterns in the hides of striped and spotted animals, especially in cats, zebras and crocodiles may be rationalized as the consequence of the distortion of the pattern by the same forces that formed the bones (Plate 44). The vertebrate hide is a dorsally compressed image of a duplicate of the embryonic patterned surface at an early stage (Plate 45). As such, the vertebrate hide pattern is a mimic of the skeleton, which it encompasses at the end of embryogenesis.

These graphic records result from the very process by which the body is fabricated. The structural plan of the living body is that of a spherical, pressurized, multilayered bag that maintains the system by adding layers internally, and removing outer layers by bursting and sloughing off, maintaining at all times, a strong, impervious pouch of five to ten layers, characteristic of the vertebrate body.

As new layers are formed interiorly, the outermost layers are compressed dorsally where they may express their patterns in adulthood. The striped and spotted vertebrates display patterns that mimic predicted flow patterns in the originally endoplasmic surface of the embryo at the end of gastrulation. Animal hide patterns may be used to confirm predicted patterns in early embryonic tissue by reverse topological extrapolation.

The patterns may best be seen in two dimensions in the hide of the animal. The stripes or spots in the pattern in the face of the flattened hide of cats may reveal its origin as a flat projection of the pole of a sphere by the bursting of a meridian and the separation of the segmented zones (Plate 46). The flow pattern in striped animals may be understood as vectored flow directed at right angles to the stripes. Stripes may occur as the bursting pattern of an outer membrane.

PLATE 44: *Leopard Hide*

Uniquely, armadillos and turtles preserve the very patterns of their primordial stages. The shell of the armadillo is in the form of the predicted articulated sphere of the egg membrane, compressed dorsally. The adult animal can roll into a sphere mimicking its pre-embryonic position. The armadillo shell is an enlargement of the mature phenotypic vertebrate pattern. It is the literal enlargement of the articulated sphere of its own egg.

The uniform pattern of turtle shells can be rationalized by a topological derivation from a hexameric sphere where five segments compress dorsally forming the carapace, leaving one flat segment as the plastron. The turtle shell may well depict the surface of the primordial oöcyte itself (Plate 47). Vertebrates slough off their outer surface layer, except for the turtle and armadillo, which retain them by compressing them dorsally.

The armadillos and turtles, as well as fish, birds, reptiles and mammals, carry on their backs graphic memoirs of their extreme infancy. These may be printed out in succession at later stages. The patterns of young animals disappear and are gradually replaced by the adult pattern. These vertebrate annual rings provide the kind of corroborative evidence for their own generative origins that crystallography provides for the existence of a crystal lattice of atoms.

The diverse axially repeated pattern of snake skins may be the result the so-called Couette-Taylor effect, which is the generation of patterns in a fluid held between counter-rotating coaxial cylinders. The lengthy cylindrical extrusion process of snake embryology could easily be subject to this effect.

PLATE 45: *The vertebrate hide is a dorsally compressed image of a duplicate of the embryonic patterned surface at an early stage. As such, the vertebrate hide pattern is a mimic of the skeleton, which it encompasses at the end of embryogenesis.*

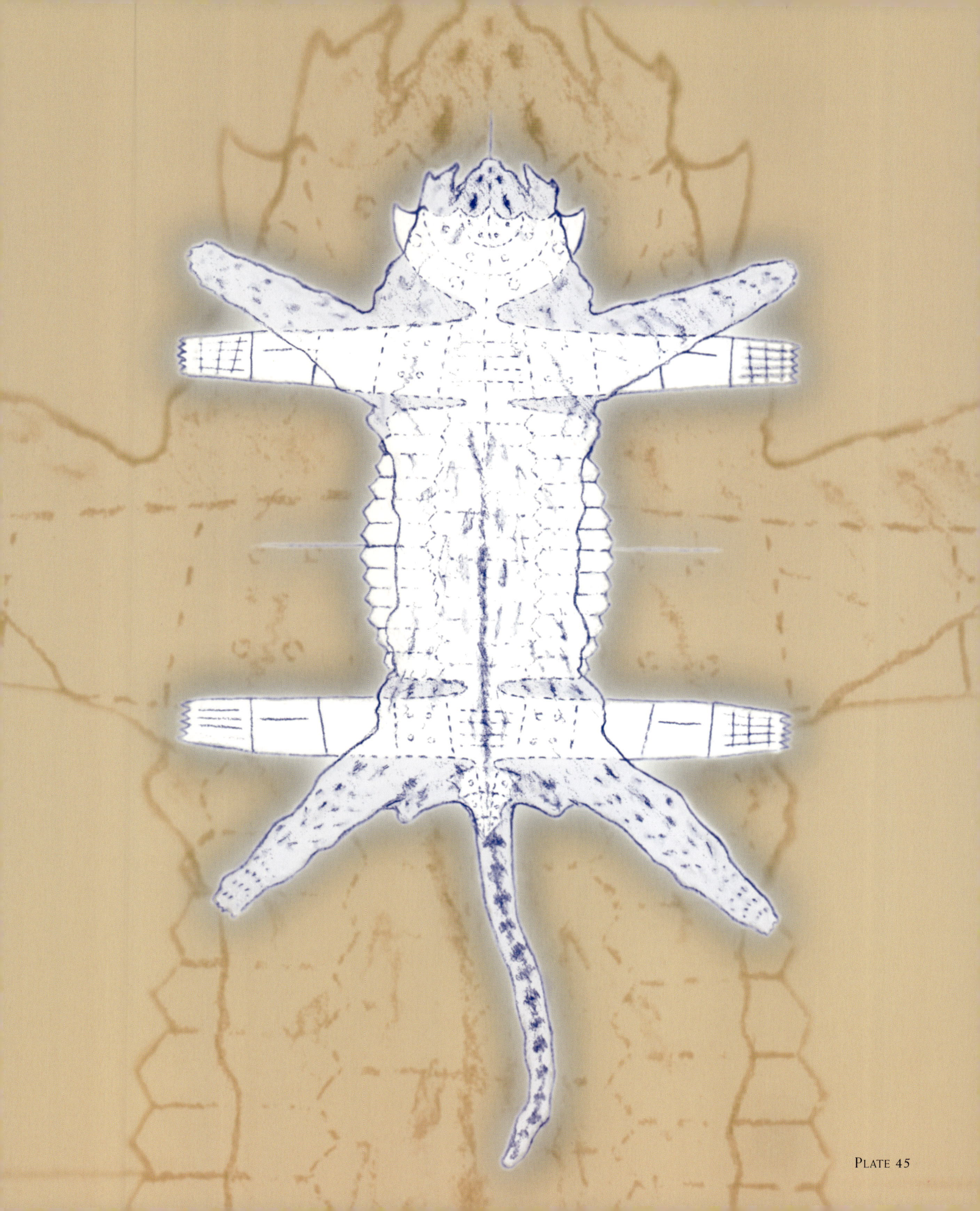

Plate 45

PLATE 46: *Cat Face. The patterns may best be seen in two dimensions in the hide of the animal. The stripes or spots in the pattern in the face of the flattened hide of cats may reveal its origin as a flat projection of the pole of a sphere by the bursting of a meridian and the separation of the segmented zones.*

Plate 46

PLATE 47: *The uniform pattern of turtle shells may be rationalized by a topological derivation from a hexameric sphere where five segments compress dorsally forming the carapace, leaving one flat segment as the plastron. The turtle shell may well depict the surface of the primordial oöcyte itself. Vertebrates slough off their outer surface layer, except for the turtle and armadillo, which retain them by compressing them dorsally.*

PLATE 47

Chapter 16 Natural Form

If forms as complex as the protein folds are intrinsic features of nature, might some of the higher architecture of life also be determined by physical law? The robustness of certain cytoplasmic forms, for example the spindle apparatus and the cell form of ciliate protozoans such as Stentor, suggests that these forms may also represent uniquely stable and energetically favored structures specified by physical law. If it does turn out that a substantial amount of higher biological form is natural, then the implications will be radical and far-reaching. It will mean that physical laws must have had a far greater role in the evolution of biological form than is generally assumed. And it will mean a return to the pre-Darwinian conception that underlying all the diversity of the life is a finite set of natural forms that will recur over and over again anywhere in the cosmos where there is carbon-based life.
(Michael Denton and Craig Marshall 2001)

In a startling scene in the Stanley Kubrik film "2001: A Space Odyssey," space explorers on an uninhabited planet encounter a huge rectilinear slab. Their first thought is that some intelligence must have preceded them, in order to explain the existence of a perfectly geometric form among the otherwise random disorder of the terrain. However, inorganic nature often produces fairly large objects of ordered symmetry in the form of huge crystals and geological formations.

Symmetry and ordered form are more the rule than the exception in nature. At the molecular, atomic and sub-atomic levels all form is perfect. Regular forms underlie the motion of particles in circular or elliptical orbits, or the outline of force fields surrounding objects with mass or charge. Nucleons and atoms associate with each other according to simple, invariable laws of attraction and repulsion. Molecules define regular forms by the configurations assumed by their atoms. Now even the folds of proteins are known to be determined by simple physical laws similar to those giving rise to crystals and atoms. There is a set of "constructional rules" governing just how these protein folds will occur. Denton and Marshal say that biological forms might be intrinsic features of nature generated by the same physical laws that govern the universe (Denton 2001).

Inorganic molecules are able to assemble themselves into naturally occurring macro-molecular forms called crystals. These are structures of perfect, regular form according to any of just six geometric patterns. Some attain sizable proportions. The gleaming monolithic obelisk in Stanley Kubrik's film could be a huge, single, perfect cube of calcite.

PLATE 48: *Lattice-like arrays of particles or hexagons have been observed on membrane surfaces created by the packing of phospholipid molecules. Upper and middle panel Müller (1997); lower panel, Stewart (2001) Copyright Ivy Press*

"Modern biology has come to occupy an extreme position in the spectrum of science, dominated by historical explanations in terms of the evolutionary adventures of genes. Physics on the other hand, has developed explanations of different levels of reality, microscopic and macroscopic, in terms of theories appropriate to these levels..."
—Brian Goodwin

Perfect forms can be created as artifacts of motion. A volcano can spew objects into the air in perfect parabolic trajectories. An object falling into water can create a symmetrical fountain display as well as concentric wave circlets on the surface. The moon's craters are perfect circles formed by explosive hypervelocity collisions of asteroids and comets. Fireworks displays are formed in this way, as we gasp at their perfection. The reason for the characteristic fluidity of organic form is that it is produced by the sudden freezing of a streaming fluid.

Organic nature is the other source of large regular forms; there are endless examples of simple symmetrical shapes assumed by growing plants and animals (Plate 48). Some viruses look like miniature spacecraft (Warner 1964; Wallace 1997). Flowers are arrays of petals as perfect as if produced in a kaleidoscope. The centers of sunflowers are as if computer generated. Single cell organisms such as radiolarians and diatoms are easily confused with inorganic crystals in their symmetry and perfection (Plate 49). The snail and other shellfish extrude structures of mathematical perfection.

In these cases, the origins of the forms can be understood and sometimes observed in the early embryological processes that stack cells one upon the other in orderly fashion. Repetition of the process will generate patterns on a fairly large scale. Similarities to process of inorganic crystallography are striking.

In both cases a simple, monomeric configuration dictates the arrangement by which the elements stack upon each other like oranges in a carton resulting in a patterned array. Such patterns may or may not have a direct relationship to the configuration of the monomer itself. Sodium and chlorine atoms associate with each other as the corners of a cube. Crystals of table salt under a lens are as perfect as dice, suggesting that the crystal is a blow-up of the molecule of which it is formed Other inorganic monomers dictate three-dimensional arrays, which are not necessarily directly related to the form of the molecule. A massive array of Ping-Pong balls is not necessarily spherical, but will otherwise embody the characteristics of a crystal.

So it is that organic life as well as inorganic nature can display the phenomena of crystallography as a source of its morphology. To extend the analogy, one can cite regular patterns occurring in more complex organic forms. Obvious examples are seen in the beautiful, orderly surface patterns of colored stripes and spots in so many animal species. Patterns become part of morphology itself in cases like the scales of fish, or the integument of turtles, armadillos and crocodiles.

Thus, to considerable extent, the morphology of living form can be understood by purely geometrical considerations. This seems to include characteristics that are noticeably regularized, patterned or symmetrical. These situations stand as exceptions to the general understanding that living morphology is dictated by the genetic material, which serves as a blueprint for the form of the organism. This reasoning leads to the conclusion that there are two separate effects that define living form: the genetic and the geometrical.

PLATE 49: *Self-organized phyllotaxis patterns appear in sunflowers and microscopic marine diatoms.*

If there is a boundary between the two, it is not yet defined. Genetic theory ascribes form to the DNA blueprint, leaving the obviously geometric forms as the exception.

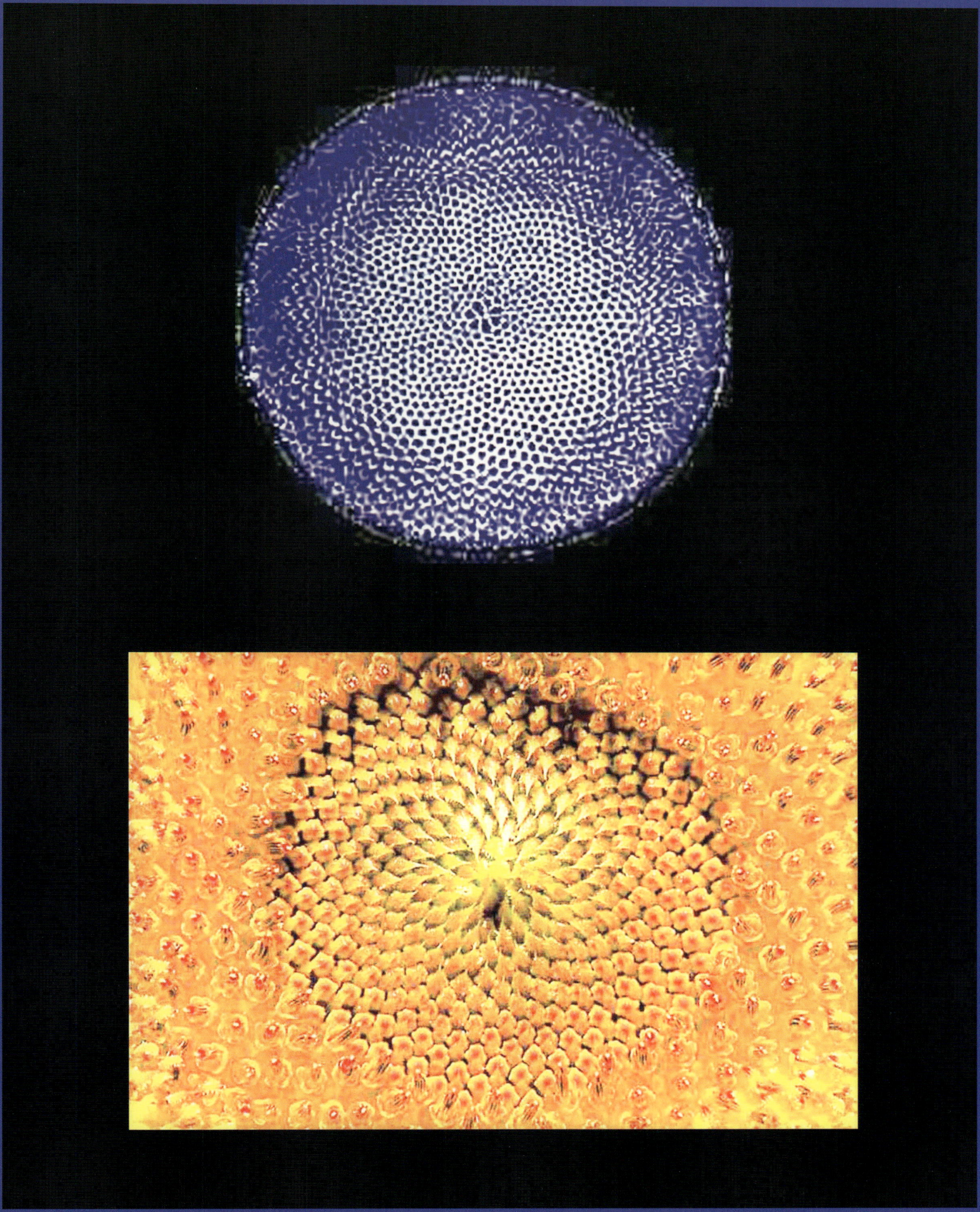

PLATE 49

Chapter 17 The Myth of Biodiversity

The agreement of so many species of animals in a common schema...grounded...in their skeletal structure...gives at least a glimmer of hope that the principal of mechanism, without which no science of nature is possible, may accomplish something here. (Immannuel Kant 1790)

In spite of what appears to be an enormous diversity of life forms, the most fundamental principles of biochemical and body plan construction are astonishingly alike. From a morphological perspective, the familiar multicellular bilateral animals, which we divide into three or four main groups, are basically all the same. These include the fish, amphibians, reptiles, birds, mammals and the arthropods. It is as though nature deliberately obscured the solution to the riddle of the origin of life forms by presenting such a bewildering array of forms as to divert thought into an unending effort to describe how they differ one from another, rather than to perceive how closely they are alike. Mathematics, physics and chemistry are organized according to simple generating principles that account for the diversity of their respective phenomena. Biology, however, in focusing on biodiversity, has made no progress in identifying a unifying theme in the shape of living form, which could lead to the common rule governing their construction.

The phyla of bilateral animals, which represent most animal life, are subdivided into millions of species in the spectacle known as biodiversity (Tudge 2000). But the diversity consists of the diversity of myriad variants on only a few basic body plans (Plate 50). While man-made objects found in the house or workplace represent thousands of truly different forms as diverse as a piano, spoon, toaster, teacup, umbrella, hat or shoe, all of the bilateral animals are variants of only one fundamental body plan. Dog, cat, lobster, hummingbird, whale, turtle, June bug, crayfish, scorpion, spider and man, have in common a semi-cylindrical bilateral body, curled in an arc, subdivided axially by around twenty segments, that terminate in paired, jointed, clawed limbs, sectioned as an anterior head, a thorax and tail; the head featuring paired eyes, a mouth beneath, connected to an interior tube exiting at the ventral posterior; the more highly developed uppermost section harder and darkly pigmented; the underneath, pale and soft (Plate 51). It is plausible that one sequence of events would be capable of accounting for the origin of the single, geometric form shared by all bilateral animals.

PLATE 50: *Homology*

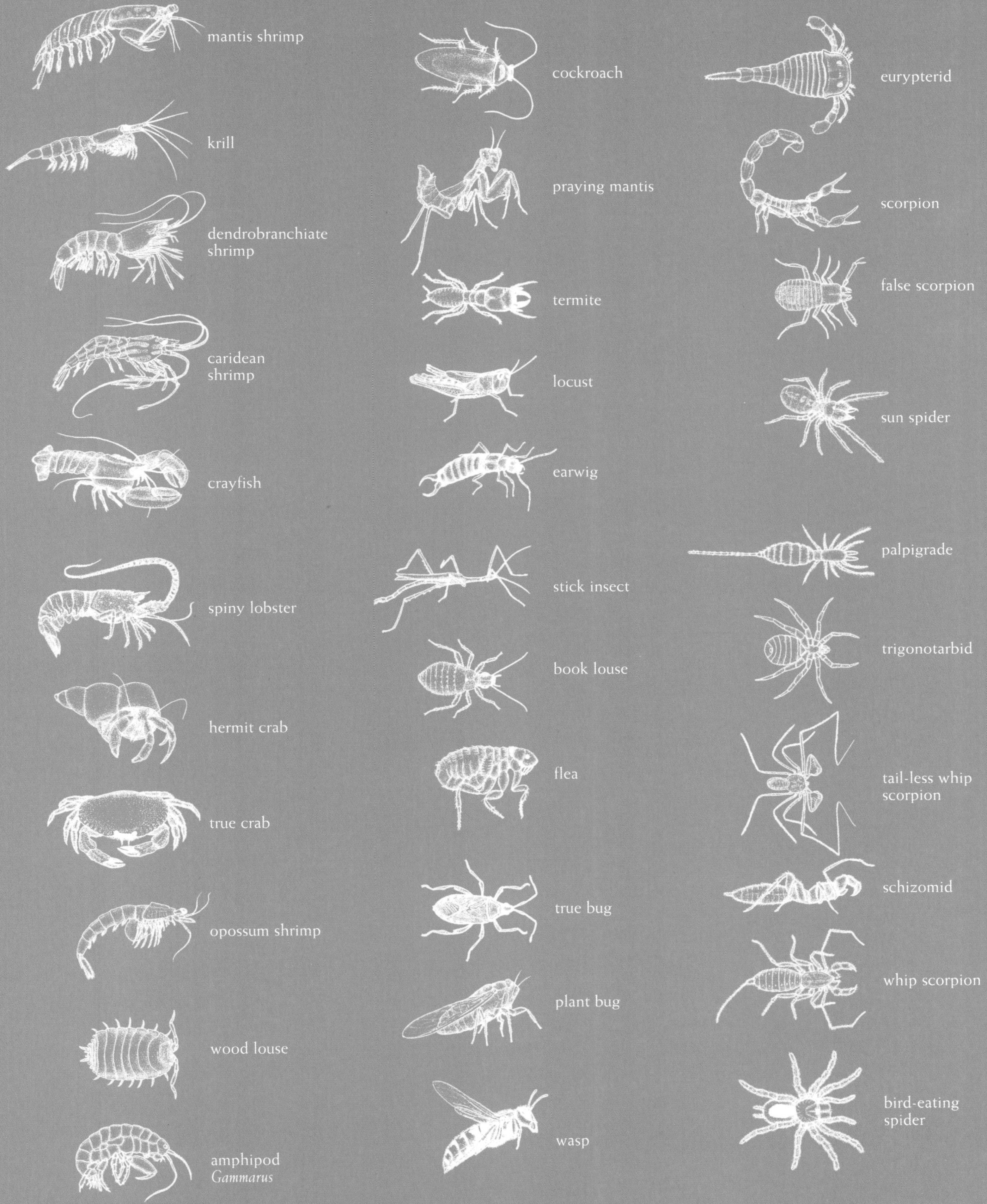

mantis shrimp
krill
dendrobranchiate shrimp
caridean shrimp
crayfish
spiny lobster
hermit crab
true crab
opossum shrimp
wood louse
amphipod *Gammarus*
cockroach
praying mantis
termite
locust
earwig
stick insect
book louse
flea
true bug
plant bug
wasp
eurypterid
scorpion
false scorpion
sun spider
palpigrade
trigonotarbid
tail-less whip scorpion
schizomid
whip scorpion
bird-eating spider

A glance at the geological landscape and the heavenly bodies shows that the material within the universe forms aggregations that assume an ample, but limited number of shapes and forms. Galaxies are spiral and elliptical; stars, planets and atoms, spherical. Geometry dictates five regular polyhedra. Crystallographers recognize some three dozen kinds of crystals. Geology describes conical volcanoes, linear faults and rifts, multilayered strata and circular craters. These forms are duplicated all over the universe, independently arising out of a uniform set of physical laws. Biologists call convergence the coincident appearance of like structures in otherwise unrelated organisms. Whether biology follows these rules in particular instances has been hotly debated.

Biologists recognize the principle of inheritance, by which siblings resemble each other and their parents. Living forms may resemble each other by common ancestry as well as by independent common evolution. Life may be classified as a single tree or as several separate bushes. From a common mammalian ancestor, the continents of Africa, America and Australia have each independently produced mammals with remarkably similar forms. A striking example is the resemblance between the recently extinct Tasmanian "wolf," which is an Australian marsupial "dog," and the placental dogs common on other continents. Although the two are very far apart on a phylogenetic tree, it takes a skilled zoologist to distinguish them by skeletal features alone (Angie 1998).

All the vertebrates and arthropods have the same body plan, that is, a segmented semi-cylinder with paired, jointed limbs, an anterior head with paired sense organs, ventral mouth, thorax and abdomen (Plate 51). However, vertebrates, insects and crustaceans are all believed to have arisen separately.

An analysis of the metabolic networks of forty-three organisms including plants, animals and prokaryotes (eubacteria and archaebacteria) by Jeong and a group of researchers (2000) showed that despite significant species differences of the individual constituents and pathways of networks, their metabolic organization is identical. Furthermore, they are also universally endowed with designed principles of robust, error-tolerant, scale-free networks. Minelli (2003) in his discussion of this work points out that universality of mechanisms does not imply that exactly the same molecules are involved. "If similar, but not identical molecules actually perform the same job in different organisms, this means that the same result was probably obtained independently more than once." Morphogenetic and patterning effects in living systems thus seems to be the inevitable outcome of physical properties of cells and tissues (Newman and Comper 1990).

Morphological resemblances are no grounds for assuming common ancestry among the phyla. The obvious resemblance of the insects, crustaceans, scorpions and spiders is insufficient reason to assume common ancestry. The famous Cuvier-St. Hilaire public debates of the 1830s, reviewed in the press by Goethe, had to do with whether, according to Cuvier, the classes of arthropods and the vertebrates were separately created by the divine or, according to St. Hilaire, descended from a common ancestor. They were both wrong, Cuvier more so.

The debate on the origin of life forms continues, often by vehement contenders. Today the arena pits natural selection against self-organization (Wesson 1994; Dubos 1980; Dobzbansky 1951). The question is whether life forms are randomly chosen, contingent on unrepeatable sequences of changes in the environment, or whether they arise from some underlying system of order, as do crystals, geological formations and the very universe. In a *Natural History* article and later *New York Times* cover article (Papineau 1995), the late Stephen Jay Gould defended natural selection against Simon Conway Morris (Morris 2003) who cites the phenomenon of convergence, the coincidental appearance of features in unrelated life forms as evidence that the appearance of humanoids is inevitable.

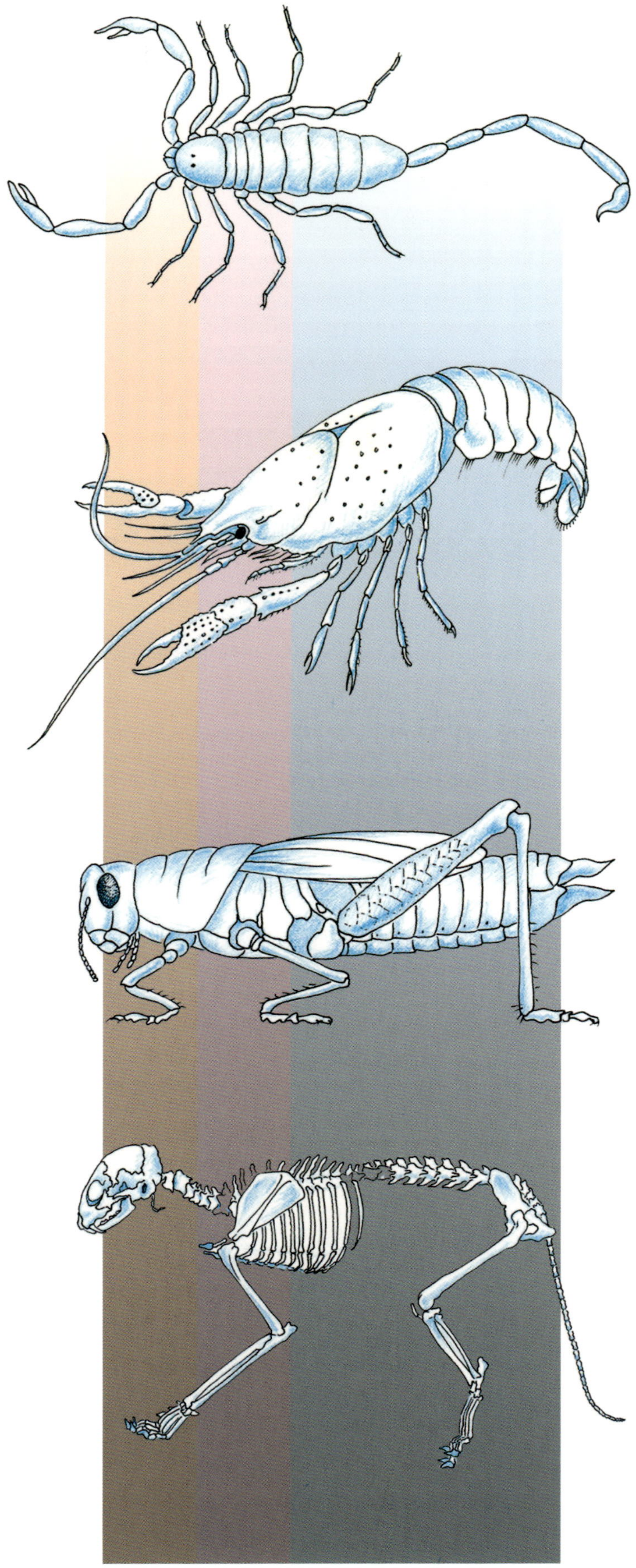

PLATE 51: *All the vertebrates and arthropods have the same body plan, that is, a segmented semi-cylinder with paired, jointed limbs, an anterior head with paired sense organs, ventral mouth, thorax and abdomen.*

Chapter 18 Life is Plastic

Innovative phenotypes arise from preexisting phenotypes...biologists not actually engaged in the study of variation may be surprised by the extreme variability observed even in vital organs such as the human heart and stomach...variations that might be considered monstrous pathologies if seen in isolation occur as part of normal development in human populations. (Anson 1951)

Often developmental plasticity refers to phenotypic plasticity when, in response to different external conditions, different phenotypes are produced (Schlichting 1998; MacDonald 1940; West-Eberhard 2003). If a grass-feeding caterpillar of the moth *Pseudaletia unipuncta* is reared on hard grass, its head grows much bigger than if it were reared on soft wheat seedlings. Larger heads are definitely better in dealing with hard grass (Bernays 1986). The tadpoles of the spade-foot frog develop into either an omnivore morph or a carnivore morph depending of food present, with two entirely different heads (Pakkasmas 2001). Emma Whitelaw, a biochemist at the University of Sydney, works with genetically identical mice with different coat colors that pass on their different colored coats to their offspring (Rakyan 2003). The list of these examples is endless. The environment is not merely a permissive factor in development; it is also instructional, and different environments can produce different phenotypes from the same genotype. "The environment may determine sexual phenotype; it may induce remarkable structural and chemical adaptations according to the season, or specific morphological changes that allow an individual to escape predation; it may induce cast determination in insects. The environment can also alter the structure of neurons and the specificity of immunocompetent cells" (Gilbert 2003). "This means that reductionism cannot provide a complete explanation of development" (Gilbert 2000).

However, plasticity is not limited to differences in response to environments. Two- and three-headed calves and turtles; human polydactylism, conjoined twins, and all manner of teratologies regularly appear in nature. These have been recorded for centuries as part of natural history. There is currently a veritable plague of frogs and other amphibians showing extra limbs and eyes. Monstrous forms can be made in the lab by tying a string around a frog egg, by exposing the developing embryo to radiation or mutagenic chemicals, or by manipulating a single gene. These always present a departure from the phenotypic anatomy by a qualitative change in an organ, or a homeotic appearance, that is, the occurrence of an organ in the wrong place.

PLATE 52: *Pig with Two Bodies*

However, neither naturally occurring nor laboratory-induced teratologies ever produce a novel body plan. Phenotypic pattern is frozen.

Conceivably each example of an anomaly is anatomically unique, having never occurred in the sequence of a species' evolutionary history (phylogeny). Each may display a complex skeleton and muscle system never before seen, including such novel forms as branched vertebrae, perfectly fitted for the new configuration. These anatomy systems did not evolve, yet they function perfectly the first time they are ever used, at least for some length of time, if not for the normal lifetime of the individual. Dogs show remarkable and unexampled morphological diversity from Chihuahua to Great Dane. Every cross, mongrel or "mutt" is morphologically distinct and possibly unique.

The genome could not possess plans for all possible accidental morphological eventualities and singularities. Does the genome direct only normal morphologies? What directs anomalous morphologies? These questions are unanswerable in the context of conventional genetic theory. One possibility is that phenotypic development is not directed by the genome.

Although we are used to individuals in animal species being alike, in plants the phenotypic form is fundamentally plastic. No two trees in an orchard have the same configuration of branches. It would be remarkable to find two alike. Roots find their way between grains of sand and pebbles. A tree will grow around a boulder. The tissue of tree burls flows like lava, acted on moment by moment only by the immediate or proximate forces. Topiary artists can produce any shape of tree.

The multi-celled organism may be thought of as basically plastic. A sheet of tissue expands in the directions dictated by the prevailing circumstances moment by moment. Cells divide along their smallest circumference, which is always perpendicular to the force causing the deformation, resulting in growth in the direction of least lateral pressure. Identical individuals within the species depend on some system to overcome an overriding tendency toward the plasticity of living organisms. Species robustness is assured in animals by the reliability of the genome, acting in the protective environment of embryogenesis. Teratological anomalies easily occur if either is disrupted (Plates 52, 53).

PLATE 53: *Courtesy, Mütter Museum, College of Physicians of Philadelphia (Worden 1984)*

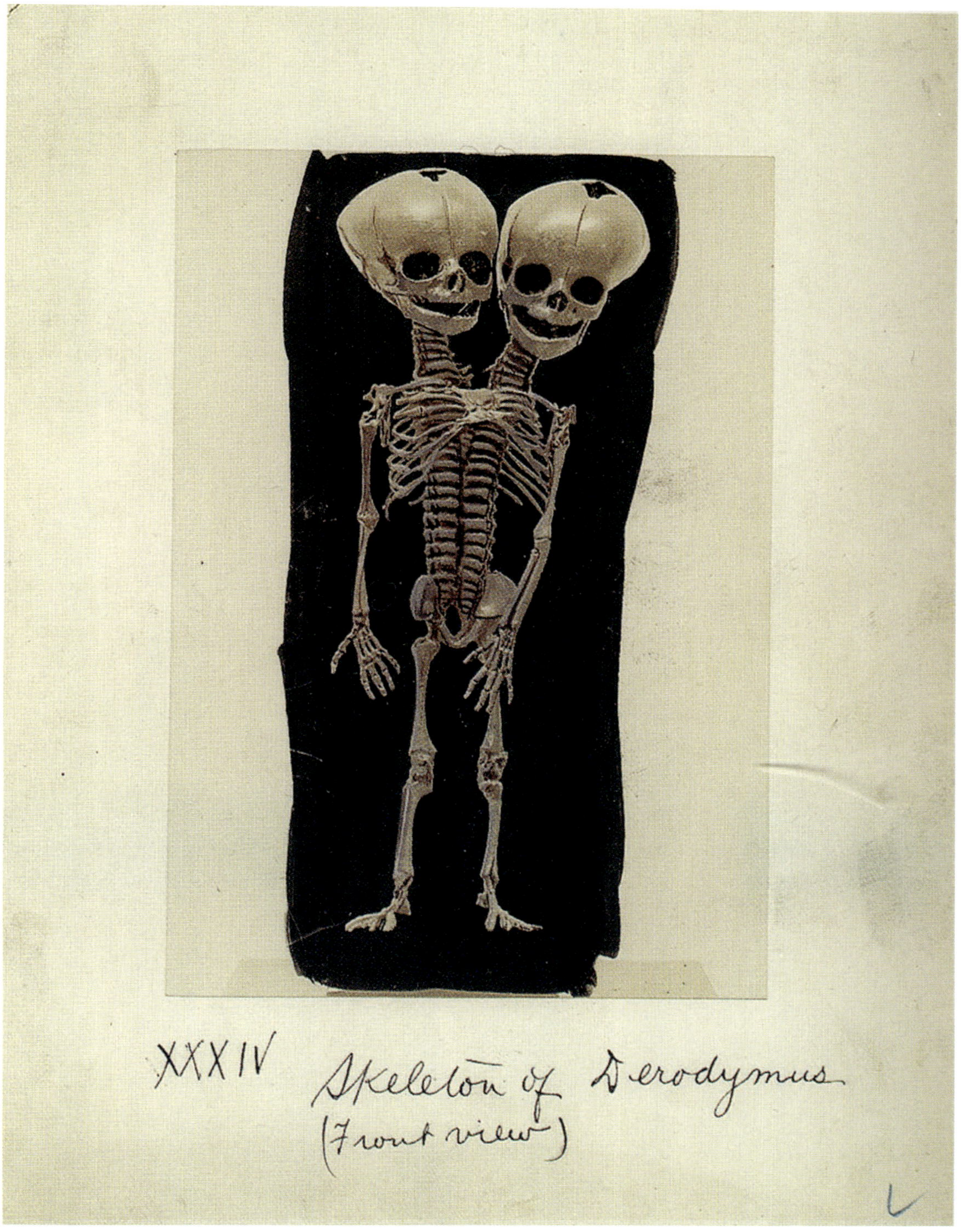

PLATE 53

Chapter 19 The Archetypal Plant

The old idea came back to me. Could I not make the discovery among this throng, of the archetypal plant, for surely there must be such thing! Otherwise how would I know this or that creation to be a plant, if they were not all formed from the same model?
—Goethe, at the sight of the flowers in the public gardens in Palermo

Goethe never fully described his vision of the archetypal plant beyond a few simple sketches (Plate 54). It remains in the same limbo as Fermat's last theorem. Plants are ideal candidates for the study of mechanical causes for growth and symmetry. There is little need to store the complex and exhaustive map of the lettuce shape in its genetic information; the elasticity of its construction can result from simple growth laws (Roman 2003).

The patterned surface of the embryo, the primordial protodermal sphere, is transformed into a stalk of coaxial cylinders each bearing the pattern. Slight changes in the registration of the adjacent cylinders produce the celebrated patterns of phyllotaxis, which by numerological, geometric coincidence demonstrate Fibonacci numbers and the golden ratio producing the beauty of a sunflower or the spiral of a pinecone. "Explaining these patterns," wrote Charles Darwin in 1859, "could drive the sanest man mad."

"In the 1990s Hox genes were much discussed in plant morphology. The differentiation of the four basic floral elements (sepals, petals, stamens and carpels) were interpreted as deriving from the homeotic control of three main classes of genes: A, B and C. A would specify sepals as the element of the most external whorl in the flowers; those of the following whorl would instead develop as petals in the flowers, stamens and carpels, under A + B, B + D and C function, respectively. It was soon realized, however, that things are more complex. At least two A function and two B function genes have been discovered in the model plant *Arabidopsis thaliana*. The ABC genes, if expressed in a part of the plant other than the flower, do not transform leaves into flower elements" (Minelli 2003).

It became such an increasingly complex picture that many researchers now wonder if it is of any use at all to try to interpret flower phenotypes strictly by molecular genetic events (Theißen 2001).

Pressure of the cell within the cell walls is high, ranging from seven to ten times atmospheric pressure. The cellulose that holds the pressure is as stiff as high-grade plastic so even a humble daisy resembles a mechanical pressure vessel (Green 1999). It was Paul Green who explained the development of plant structures in terms of buckling of the plant skin. Patterns form at a growing tip, the meristem, which resembles a pressurized dome. Elements of the pattern or primordia emerge near the edge of the meristem where mechanical calculations show that the skin of the meristem is compressed, not stretched by the underlying pressure (Brookbank 1978). Green and his group hypothesized that the compression would cause the meristem to buckle and spur the growth of the primordia. Using computer simulations, they showed such buckling could create wrinkles that mimic the Fibonacci patterns of sunflowers (Green 1999). Green looked at the plant cell as the mechanical instability of a shell, similar to what patterngenesis proposes for the oöcyte. He found a linear relation between the epidermis thickness and the wavelength between the development of new leaves. This relationship was similar to the buckling wavelength calculated from the geometry of the epidermis and the interior cells (Steele 2003).

Toroidal Topology in Flowers and Fruits:

With growth the exo-surface is forced to enter the interior at a pole, where it grows internally as a shaft of concentric cylinders. This shaft penetrates the spherical mass, emerging at the opposite pole. Each of the concentric circles subdivides the region around the pole producing a net of coaxial cylinders, each bearing the empatterned surface of the beginning sphere. The shaft then consists of layers of tissue each bearing a cylindrical transformation of the early primordial pattern. These layers now superimpose on each other, slightly out of register, a geometrical conclusion of the beginning pattern.

Beginning at the apex, entering and extending through the plant body, axial growth of the stalk causes a cylindrical shoot, the apical meristem, to extend upward. This is constructed of cylindrical sheets of tissue originating at the apical annulus and thus bearing the imprint of the embryogenic pattern transformed into cylindrical form. Morphological order is thus conveyed to the pattern to the stem by this topological transformation. The stem consists of a bundle of tubes enclosed in a two-layered outer cylinder. Periodic swellings generate the form of the flower. Internal tubes form the carpel system consisting of a central ovary surmounted by the stigma, surrounded by the group of tubular male organs surmounted by the anthers. An enclosing stem bi-layer produces the corolla of petals, supported by the supporting outer structure, the calyx. The origin of these forms in a homunculoidal sickle torus is described in the first chapter.

Symmetry in Plants (1994a, 1994b), the definitive work on plant morphology by the great botanist Roger Jean, establishes the basis for a theory of plant morphogenesis independent of the genome. Structural periodicity is the name given the phenomenon of the ubiquity of patterns in plants and mineral crystals. Phyllotaxis, the basis for plant form, is ostensibly a geometric system.

Shapes of leaves, calyx and petals may be understood as depending on the fracture pattern of the sheet from which they are formed, as it bursts from the growth of the layer beneath, exactly the same way each time.

PLATE 54: *Goethe's vision of the Urpflanz, the archetypal plant.*

One of the most impressive achievements in botany, during the last hundred years, has been the accumulation of data bearing on the mode of ontogeny from the fertilized egg, and also the development of the shoot from the meristematic apex. The most recent work tends to show that these two phases in the development history cannot be isolated from one another, since the organization of the mature shoot is merely a modification of the original 'pattern' in the relevant part of the embryo. This holistic conception of plant ontogeny is irreconcilable with a claim that has sometimes been made, or implied, that a causal explanation of plant form is to be sought in the developmental history. This claim involves a conception of development as consisting of stages which are discrete entities, each stage being treated as the cause of that which follows it, and as being itself caused by that which precedes it; but this is to confuse the succession in time of a continuous series, with causation. Two phases may follow one another as the night the day, but this does not entitle us to call the first the cause of the second. It is necessary for the biologist to achieve the difficult task of realizing the individual organism as consisting of a unification of every phase of its existence from the fertilized ovum onwards; it is a fatal error to see the individual as a summation of stages. These stages are not realities; they are the imaginary result of cutting a continuous series into segments, and then stringing them out in time. The human mind goes through this process for its own convenience, and then forgetting the artificiality of the situation which it has created, it proceeds to endow its own constructions with a causal relation to one another. If then, we set aside the hope of finding the causes of plant forms in developmental history, we may yet detect therein indications as to the manner in which some of the features of this form have come into being. It is clear, for instance, that phyllotaxis, and the numerical relations of the parts of the flower, depend upon the rhythmic development of the primordia at the growing apex. (Arber 1950)

PLATE 55: *Schematic Diagram of Plant Stem Morphogenesis*

PLATE 55

Chapter 20 Morphogenic Fields and the Cascade Theory

Genes are known to trigger the formation of an organ, but the actual shaping of the organ by genes has not been observed. "It is as if we had an architect's blueprint for a house, but never actually saw a construction crew in the act of building it." (McGinnis 1984) The current, standard explanation of the system by which the genes create the shape of the organ is sometimes explained by biologists as the Cascade Theory. According to the theory, the gene activates other genes, which activate still others in a hierarchy of genetic activity, which eventually carves out the shape of the organ in a system so complex that it may never be sorted out. This has never been demonstrated.

Darwin proposed that today's species evolved from ancient forms by the natural selection of favorable mutations, similar to the breeding of domesticated animals for favored traits (Muller 1869; Milner 1993). His followers extended the idea to say that the first forms arose this way. The modern science of genetics claims that the individual is formed by a complex, inherited code, which directs the assembly of the cells as they proliferate, and that the code resulted over eons from the preservation of random errors, which produced favorable mutants, selected by the environment.

But most evolutionary biologists today reject phylogenesis by natural selection as implausible. While no system has been demonstrated heretofore as a substitute, morphogenesis by self-organizing patterns is widely considered as a candidate since it was first proposed by the great English mathematician Alan Turing (Saunders 1992; Turing 1952a; 1952b). He developed a series of highly complex theories whereby systems of chemicals reacting together and diffusing through tissue can create spontaneous patterns. He called these elements morphogens, "shape creators." His work was based on the observation that certain chemical reactions generate wave-like patterns in the substrate medium, notably the famous Belousov-Zhabotinskii reaction, an inorganic chemical reaction conducted on a flat plate where the two reagents generate wave patterns as they combine. Most recently Complexity and Chaos Theories suggest the origin of form is in the self-organization of small elements.

The genetic theory of morphogenesis is based on the phenomenon called Morphogenic Fields, observed gradient fields of growth factors. These are purported to create the final form of the organ by stimulating cell development in a two-dimensional grid field of gradients in the embryonic tissue, which invisibly prints out the shape like a figure on a TV screen. There is indeed a gradient field, and indeed a limb does appear, but no indication of the final form can be observed in the field itself. It is a proposed but undemonstrated mechanism, which could conceivably work. But the appearance of morphogenic fields may be explained in another way: The observed chemical gradients in the morphogenic field may be the result, rather than the cause, of morphogenesis. They may well be merely the chemical traces left by the RNA growth factors, which act as switches in the manufacture of protein during the development of the shape of the organ. These could be guided by some other means, for example, the mechanical exigencies inherent in the proliferation of the cells during the formation of the organ, including the effect of mechanical force on gene expression. As noted in an earlier chapter, morphogenesis is better explained under the assumption that the genes are formed by the self-organized growth patterns of the cells rather than that the genes direct the way the cells organize. Morphogenic fields consist of traces of gene product left during the formation of the organ, rather than a chemical field generated to guide its growth.

Chapter 21 Scientific Dissent from Darwinism

It is now approximately a half a century since the neo-Darwinism synthesis was formulated. A great deal of research has been carried on within the paradigm it defines. Yet the successes of the theory are limited to the minutiae of evolution, such as the adaptive change in coloration of moths; while it has remarkably little to say on the questions which interest us the most, that is how there came to be moths in the first place. (Ho and Saunders 1979)

In one of her many talks, Lyn Margulis, asks her audience to name a single, unambiguous example of the formation of a new species by the accumulation of mutations insisting that neo-Darwinism is in a complete "funk" (Mann 1991; Margulis 2000, 2002).

The premise that Darwin was wrong will not seem implausible to those who follow current biological thought with respect to evolution and development and know that the fundamental phenomena of biology remain a mystery.

The whole issue of DNA and heredity is unresolved, a mystery remains and the mystery is embedded in the problem of morphologic evolution; the process that gives animals and their organs their functional shapes and that yields the most exquisite product of all, the human phenotype. (Edelman 1988)

Darwin himself did not have a genetic program for development, no program that guides the system through its morphogenetic transitions...the genes set the parameter values meaning that they produce component parts of the system within a wide range of values. The morphological transitions then are the consequences of the cycle of dynamics generating geometry and geometry modifying dynamics. (Goodwin 1994)

Brian Goodwin's and Stuart Kauffman's position is the most recent expression of an intellectual tradition with roots in the eighteenth-century Enlightenment. Known generically as Rational Morphology, with a line of distinguished scholars from Kant and Goethe and running through Geoffroy Saint-Hilaire, Baron Georges Cuvier, William Bateson, Richard Owen, Hans Dreisch, D'Arcy Wentworth Thompson, Waddington, and others, there has always been search for "laws of form" that would explain the striking patterns of order seen in nature (Cuvier 1834; Buffon 1749–1767; Lewin 1999; Waddington 1956).

Many prominent developmental and evolutionary biologists are more blunt, saying, "Darwinism explains everything, therefore nothing" (Rosen 1984). They no longer ascribe the cause of evolution to natural selection, nor the form of the embryo body to the action of the genes. It is the variation among organisms that is the material basis on which evolution acts (Lewontin 1998; 2001).

I am not denying natural selection, I'm saying that it does not explain the origins of biological form, of the pervasive order we see out there. On a scale of one to ten, I would rate the importance of natural selection in the context of the generation of form as close to one. (Gardner 1992)

There is a current body of belief that self-organization is the basis for organic form. Mechanisms for the self-organization of tissue and cells have been proposed since the nineteenth century. Biologists in the 1930s demonstrated the phenomenon of self-organization, suggesting that the phyletic body is formed directly from the cytoplasm of the egg cell, with no inherited information beyond the physical properties of protoplasm. Mathematicians have demonstrated and illustrated how form can emerge from chaos, ever since Alan Turing's suggestion in the 1950s of wave-propagating chemical reactions as a source of organic patterns. Turing goes so far as to say: "You don't need to put a gene in for patterning, patterns form spontaneously through the operation of the laws of physics and chemistry." (Turing 1952)

PLATE 56: *Charles Darwin (drawing from SignLabs)*

Physics imposes constraints on what biology can do.... Life is perpetually tinkering with its own mathematical basis, nudging its development in preferred directions and away from undesirable ones. (*Stewart 1998*)

Muller and Newman (2003) point out in their book *Origination of the Organismal Form* that:

The forces driving morphological evolution certainly include natural selection, the appearance of specific, phenotypic elements of construction must be taken as being caused by natural selection: selection can only work on what already exists. Again, Darwin acknowledged this himself in the first edition of The Origin of the Species.

The extension of the reasoning of mutation and natural selection has been used to explain the sudden occurrence of the Cambrian fauna in a short fifty million years, consisting of fully developed animals in three dozen phyla. But this explosion of life was built on the same bilateral paradigm, with no trace of the myriad forms nature would have had to try, according to natural selection, before settling on the head, thorax, tail, paired jointed limbs, anterior paired eyes and midline ventral mouth common to all the Cambrian animals. After the establishment of these thirty-five original phyla, from whatever cause, natural selection may well have played a role in choosing which of the possible variations on these forms will survive to constitute today's illusion of biodiversity (Newman 2003; Morris 1989).

"Although the world must have contained many new unoccupied niches during the period of the Cambrian explosion, this alone cannot account for the rapid profusion of body plans once multicellularity was established, nor for the particular forms these plans assumed" (Newman 2003). According to Newman, internal, programmed bases may have existed for form throughout time; the templates for these forms must have been established by conditional organism-environment interactions early in the history of the Metazoa. He says, "Stated simply, tissue forms emerged early and abruptly because they were physically inevitable—they were not acquired incrementally through cycles of random genetic change followed by selection. What was acquired though these classic Darwinian means were stabilizing mechanisms, which eventually moved biological morphogenesis beyond the realm of determination by generic physical processes...metazoan organisms look the way they do in a general sense, not because of specific evolved adaptations, but because of the combined effects of the various physical properties that were generic to the earliest multicellular aggregated, presumed to be chemically excitable, visoelastic 'soft matter'" (Newman 2003). Newman also believes that the physical properties of the multilayered, hollow segmental forms were virtually inevitable early in the history of metazoan life, and that it was these same physical processes that decreed the formation of organs and appendages consisting of lobes, lobules, tubules, sacs, nodules and rods.

This extremely contentious subject cannot be understood without understanding the two uses of the word evolution. Briefly restated: Darwin spoke of superficial changes such as the shape of beaks in finches, which may easily be understood as driven by natural selection. This is called microevolution, which indeed could play a role in evolution from the first thirty-five phyla of Cambrian animals established

500 million years ago, to the biodiversity of today. The term macroevolution is applied to the phenomenon of how the complex animals were formed in the first place, presumably from single cell life. Darwin did not directly state that phylogenesis was caused by natural selection, but he did not object when naturalists made this extrapolation, desperate for a mechanism of phylogenesis to support their own theories of epigenetic morphogenesis; in particular, the Biogenetic Law of Haeckel (Haeckel 1896).

The microevolution effect of Darwinism, or natural selection, cannot be doubted, but the choices it makes are not creative of form. They decide only one of two possible versions of preexisting forms. Mendelian driven microevolution chooses between a large and small version, or a dark or light version of what is already there.

But macroevolution is now generally seen as implausible as a random process. Creationists use the terms "evolution" and "Darwinism" interchangeably, misappropriating the fall of Darwin's explanation of macroevolution as proof that evolution never happened. There is currently no alternative theory of macroevolution available to satisfy the demand.

Natural selection plays only a superficial role in the evolution of preexisting organisms consisting merely of scalar and chemical changes. It may well be responsible for the length of the giraffe's neck, but not for the neck or the giraffe. Importantly, natural selection maintains morphological integrity, preserving symmetry by weeding out lopsided mutant individuals. The application of natural selection by Darwin's followers is a black box explanation of phylogenesis that today universally directs biological thought. Underlying the official use of that term is the irony of its implications.

The fundamental idea of adaptation, that organisms are shaped by their environment, is an illusion when applied to gross morphology. Natural selection indeed weeds out dysfunctional individuals who cannot compete to survive long enough to reproduce. This is not to say that it generates the form of the successes. The very fact of biodiversity, that so many millions of variants can all survive, implies that any form whatever, the large, the small, the fast, the slow, the soft, the hard, the fast or slow reproducer, furry, feathered, scaled or bald all survive. How can all features be essential to survival?

Darwinism covers up the embarrassment of a lack of a plausible account for embryogenesis, the system that organizes form, the basis of evolution. The continued use of this point by creationists has led certain scientists, fearing the power of irrational fundamentalism, to suggest that the questioning of Darwinism not be made public, not, at least, until something else is in place. This may explain the conspicuous acclaim of Darwin by many prominent scientists. They are, however, usually referring to microevolution for which it may well be valid; it is irrelevant to macroevolution.

Darwin's philosophical contribution is inestimable. His over-simplified idea that nature is like an animal breeder was immediately comprehended and accepted by a majority, which previously held with creationism, despite a century of enlightenment. One myth was exchanged for another.

Chapter 22 Evolution

A striped tiger is better hidden among the vertical striations of the long grass...and an insect patterned to resemble a flower is at less risk from predators.... A spotted leopard can merge easily with its surroundings and who can doubt the design of a walking stick or a praying mantis. A Darwinian explanation of evolution says nothing about this—it mere suggests that once there, these patterns will stay because the animal is better off with them. Can it be that evolution found its tortuous way to the striped zebra via a slow gradual and random accumulation of mutations towards stripe-giving genes? What, then, were the intermediate stages? Were their proto-zebras that exhibited some other kinds of pattern? (Ball 1999)

The term evolution is ambiguous. It is widely used in reference to the establishment of species by changes deriving from preexisting fully developed species, as in the sequence from fish to amphibian, reptile, mammal, primate, to man. The establishment of the thirty-six phyla of the Cambrian fauna, the first observed fully developed radial and bilateral animals, five hundred million years ago, from which succeeding paleontological history preceded, must have begun with a single cell. While this sequence, too, must be described as evolution, the process that established fully developed animals might well be different from the process that modified them into today's biodiversity. The predecessors of the Cambrian animals, being soft bodied, left no traces (Jahn 1944; Buchsbaum 1987). Even the progress of single cell prokaryotes, bacteria and archaea with no nuclei to the more complex eukaryotes is evolution. The development of the first metazoans from a complex single cell protozoan is yet another process that may be termed evolution.

Natural selection is the favorite and the most well-known process responsible for evolution. An unfit organism, which fails to survive until reproduction, will lead to the elimination of that strain in the population. Natural selection indeed must have always maintained editorial sway over inheritance. But Darwinian natural selection states that the very shapes of the individual and of the species result from the selection of random mutations. Many authors have long rejected this proposition as implausible, while accepting the editorial effect of natural selection without hesitation.

Selection implies that each organ as well as the organism as a whole possesses survival value by dint of some functional advantage. To attribute differential advantage, purpose or survival value to every organ of every organism implies that innumerable other versions must have once been tried and failed. The record shows the opposite. For example, many versions of the same animal survive, large and small, fast and slow, fierce and timid, inconspicuous and conspicuous. Although shells seem to offer protection, only mollusks have them, and some do not. Turtles and armadillos are the only vertebrates with shells (Kingsley 1912). All animals are prey to others despite their seeming protections. Bilateral animals have some version of jointed articulated limbs. Most seem poorly designed for speed, yet they all work. Snakes manage quite well without any limbs at all. Only spiders have more than one pair of eyes and spin webs. Certain insects make cocoons. Most do not. Only insects, birds and bats fly.

Nature finds a use for whatever form occurs. Purpose is an illusion, called the teleological fallacy. Forms have no essential survival purpose. Birds flew because their limbs became light and feathery. Some never flew. We would probably be just as well off without hair, toes or external ears, or fewer fingers. We can't run well because of our upright gait. Animals of half our weight can overcome us, and we die if exposed to the environment. Only our brain makes up for this gross design failure.

Standard arguments against natural selection have pointed out that if survival depended on fitness there would be fewer rather than more species variants; that if form was created by the selection of those feature variants best adapted to the environment, there would be fewer rather than more, yet there have been tens of millions. Also, that natural selection reasoning is circular, the fit being defined as those that have already been seen to survive. Racing cars, on the other hand, are practically identical. So are bicycles. Behavior will be seen as the mechanical consequence of the fortuitous structural relation between the sense organs, the nerves and the muscles. Most human actions can be seen as innate, as they are in other animals.

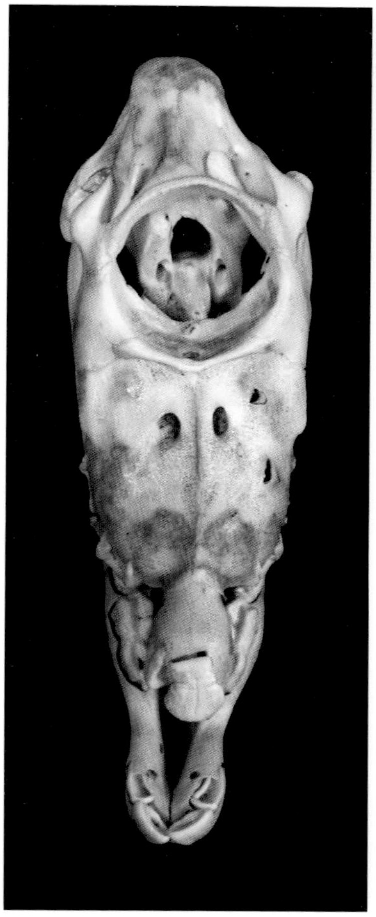

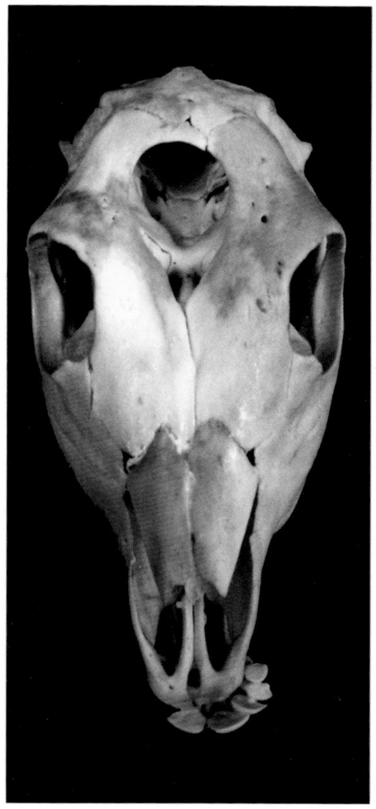

PLATE 57: *Skulls of a one-eyed horse and a three-eyed goat.*

Chapter 23 Historical Accounts of Evolution

A number of mechanisms other than natural selection have been proposed to account for phylogeny. The temptation to accept the highly plausible theory of Lamarck of the inheritance of traits (1835) derived during the lifetime of the individual has been refuted time and time again by conclusive experiments (Cuvier 1835; Giard 1907; LøvTrup 1984). After thousands of years, Jews still need to complete the covenant of Moses by surgery.

The nineteenth-century biologist Von Baer observed that in the earliest stages of development diverse animal species resemble each other. The embryos of fish, frogs, birds, turtles, dogs and humans are almost indistinguishable, differentiating only with growth (Haeckel 1896) (Plate 58). The postulate followed that nature creates each individual in recapitulation of the process of phylogeny, the development over time, beginning with a single cell, of ever more complex configurations of cells and organisms, each generation adding an additional step. This has been expressed in the epigram "Ontogeny Recapitulates Phylogeny" by Ernst Haeckel (1896), as part of the ABCs of biology, but later alternately ridiculed and resuscitated by succeeding generations of biological thought.

By the 1930s, many leading biologists endorsed endocrinal change during embryology as the prime mover of evolution and the establishment of new species. The phenomenon is called Heterochrony. An impressive list of evolutionary biologists, including Garstang (1922, 1929), de Beer (1930, 1940), Haldane (1932), Gould (1977), and Raff (1996), concluded that endocrinal changes occur in the embryological and fetal development of individuals, directed by mutation in the system of regulatory genes in the genomes. These mutations cause changes in the timing of the sequences of events in development, which result in change in the relative size of the existing organs. These changes generally relate to the difference in the pace of the separate, independent phenomena, growth and development. The pace of development relative to growth may be accelerated or retarded, producing the effects called respectively, acceleration and neoteny. Although acceleration and neoteny are the result of mutations in the genome, their effects are upon previously established organs and organisms, the forms of which may have been derived by other means.

PLATE 58: *Haeckel's famous drawing expressing ontogeny recapitulating phylogeny or that the early embryos of fish, frogs, birds, turtles, dogs and humans are almost indistinguishable.*

In his epic work of 1977, *Ontogeny and Phylogeny*, Stephen Jay Gould names the principle of heterochrony as the prime cause of macroevolution and calls neoteny the

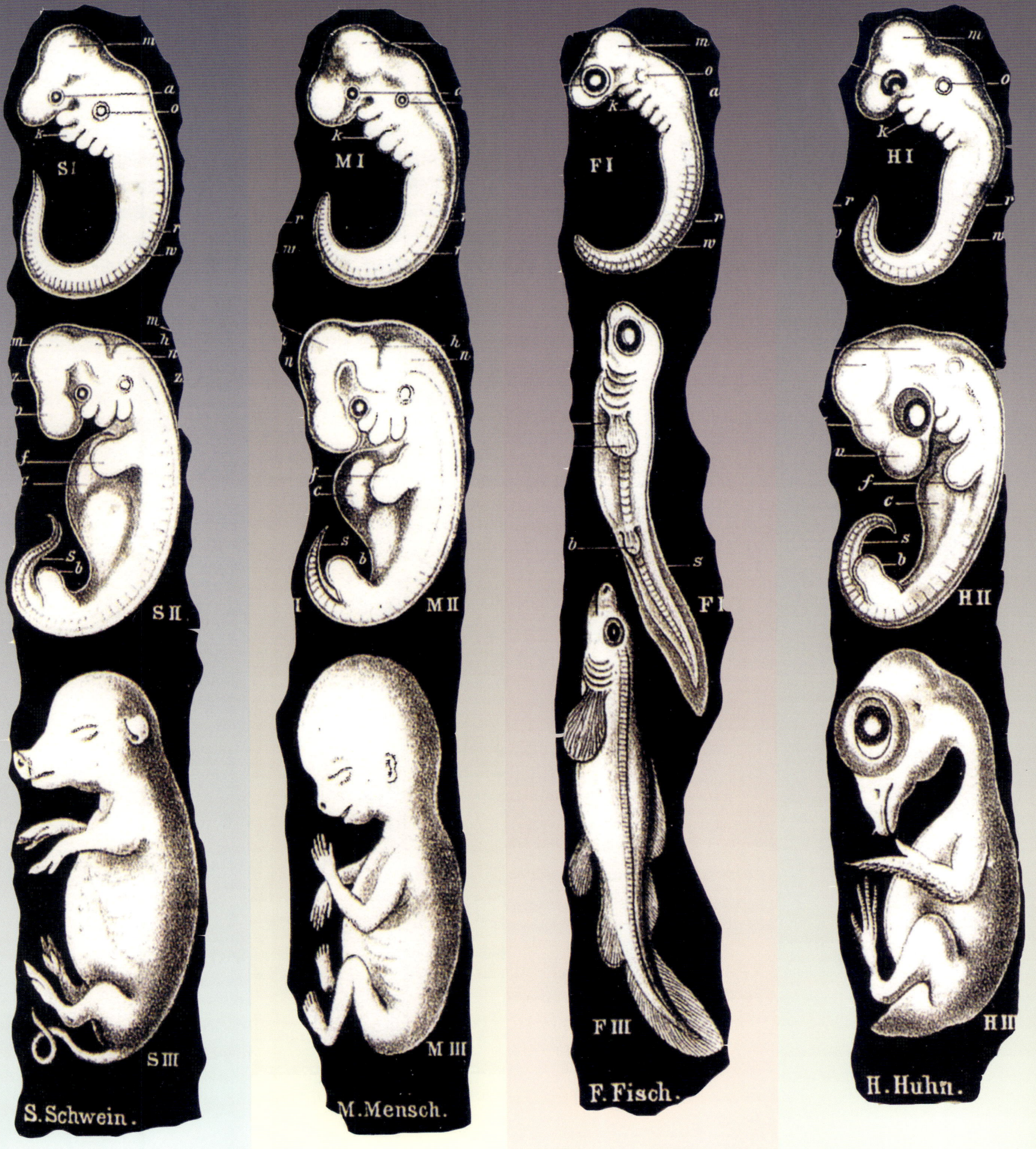
S I
M I
F I
H I
S II
M II
H II
S III
M III
F III
S. Schwein.
M. Mensch.
F. Fisch.
H. Huhn.

PLATE 58

major determinant in human evolution. In today's books and museum exhibitions purporting to represent human evolution, mention is almost never made of this phenomenon, which alone can account for virtually all the characteristics, that distinguish humans from their fellow animals. Biology texts relegate mention of neoteny to the effect on certain amphibians that can alternate between the use of gills and lungs.

With the arrival of orangutans in the Paris zoo in the nineteenth century, it was noticed that the faces of the infant apes resembled adult humans, and that the resemblance disappeared as they grew. This led to the notion that humans were juvenalized primates, first expressed by psychologist Havelock Ellis in the 1890s.

The Dutch embryologist F. Louis Bolk in 1922 established his theory of neoteny, which attributed virtually all characters, that define humanness as derivative of the effect of the single phenomenon of retardation of embryological and fetal development. The short list of human feature includes our large brain, upright posture, forward vision due to the support of the head at the bottom of the skull rather than the back, plantigrade feet, flat, small-jawed ventral face, hairlessness, longevity, ventral disposition of the vagina permitting appositive coital posture, lengthy childhood, playfulness, creativity and freedom from innate behavior patterns (Conklin 1927; Colbert 1991; De Beer 1971).

The principle underlying the phenomenon is based on the possibility of mutations in the regulatory genes of the genome, which alter the timing and intensity of the events of prenatal ontogeny. Simply put, prenatal development consists of the parallel effects of growth and development. If growth outpaces development, the individual of the species is born as an oversized fetus. Features characteristic of adulthood do not have time to manifest themselves. The effect is very obvious in humans, and may be applied to the evolution of all primates from mammals. The evolution of vertebrates from fishes through amphibians, reptiles and mammals may be attributed to the general effect of neoteny, indicating that evolution is directed by the vagaries of endocrinal changes in the sequences in prenatal development established in the first set of thirty-six phyla that appeared in the Cambrian era (Randinsky 1987; Mathew 1915; Chalene 1990; Carroll 1988).

The development of fins in fish begins embryologically with short, stubby bones, which elongate into the thin spines of fins (Holland 1995, 1999). The elimination of the terminal stages produces the walking limbs seen in the mudfish, lungfish and catfish, transient stages leading to the amphibian limb, substantially unchanged through evolution to the mammal form (Phillips 1975). Lungs develop from embryonic cavities, while late-forming gills disappear, remaining vestigially in human embryos. The vertebrate embryo begins with a relatively large head, becoming proportionally smaller toward adulthood (Guyenot 1929; Ballard 1964). In vertebrate evolution the head gradually becomes relatively larger from fish to man, as later stages are lost to neoteny (Wagner 2000).

The evolution of the egg from fish to mammal consists of a gradual reduction of the amount of volume occupied by the yolk. In fish, reptiles and birds the yolk occupies most of the space, confining the developing embryo to a flat disc at the top (Patton 1951; Gillmore 1973). The limbs of these classes are horizontally disposed, the spinal column is straight, and the rib cages generally do not come together ventrally. The mammalian egg has little yolk, permitting a more spherical configuration, including an arched spine, ventrally conjoined ribs and the vertically downward directed disposition of the limbs.

While the general trend in vertebrate evolution is toward juvenilization, the process is reversed from time to time in the opposite manifestation of heterochrony, called acceleration. For example, while the amphibians lose the highly developed integument of scales in fish, these are regained in reptiles and even progress further in birds, as the feather is the further development of the scale. While mammalian fur is underdeveloped reptilian scales, by acceleration the armadillo and pangolin regain their scaly armor.

The minor morphological differences in mammals may be accounted for by the hyperactivity of a regulatory gene at some point along the length of the genome, causing hypertrophic growth in the hindquarters or forequarters. In birds the exaggerated growth of the pectoral muscles in the confines of the egg, crowds the neck into the typical s shape, and flattens the beak in ducks and geese. The practical contingencies of development with an ellipsoidal egg can account for much of the morphology of the adult animal.

Punctuated Equilibrium, the Gould-Eldridge Theory

A main premise of Darwinian natural selection is that evolution is the gradual transformation of forms over eons, trying all possibilities along the way, as the changing environment selected only the fittest to survive. But the fossil record does not record any such thing. Species occur, remain unchanged for extremely long periods of time and then go extinct as similar ones rise along the evolutionary time, perhaps over just a few generations. Darwin expected intermediate forms to be eventually discovered. They never were.

In an analogy, the development of a modern city of skyscrapers from simple huts may be seen as the gradual increase in height of the buildings. But in fact a given structure is razed in a few days, or even a few seconds, and replaced by a taller one in a few months, which may last for a century or two. A time-lapse photography series made of a view of Manhattan, one exposure a decade, starting in 1620, would appear to show buildings growing like blades of grass.

The mechanism for the rise of new species may originate in extreme conditions at the edge of the environment, where a population of successful mutants may occur in relatively short time, before invading the interior of the territory at the expense of the indigenous population. Gould perceived this phenomenon in his studies of populations of snails along strips of tropical beaches. He discovered that different species populated strips of beach a few hundred meters long, separated by narrow bands of populations of mutants.

Punctuated equilibrium implies that new species develop almost immediately, presumably without benefit of natural selection.

Chapter 24 A History of Anatomy

The goal of theoretical biology is to solve the most perplexing problem that has confronted science since Aristotle: how one cell produces a complex body composed of trillions of cells, and how over time, simple forms achieve their present complexity. Following the tradition of Euclid, that is, the development of the complex from the simple by logical steps, each supporting the next and supported by the ones before, authors have historically sought to unravel the riddle of morphogenesis and evolution by applying mathematical and physical principles to biology as they have in mathematics, physics and chemistry with great success (Vogt 1851).

Aristotle used the word "entelechy" to describe a motivating force that, guided by transcendental principles, directed an animal or vegetable soul to generate the organism. Even into the twentieth century the term "vitalism" was used as a catch-all for an unknown force, that directs the stages of embryogenesis, which could be observed, but not understood. But the enlightening influence of eighteenth-century philosophy inspired the steadfast view that the body was a machine, its formation pursuant to the laws of physics, described by mathematics and geometry. Goethe was guided by an obsessive mission to discover the universal factor that generates all organic form. In a letter to his sister he says *Tell Herder that I am extremely close to the secret of the generation and organization of plants, and that it is the simplest thing that one can imagine... The same law will lend itself to all other living things.* This led to his famous thesis, which revealed that flower parts are derivatives of leaves.

The invention of the microscope in the seventeenth century directed immediate attention to sperm and eggs. In 1683, Leeuwenhoek saw the prefigured Homunculus in the sperm (Plate 2). Others saw the little man in the egg. Both ovists and spermists confirmed preformation. The theory of emboitment, like Russian dolls, Homunculus within Homunculus within Homunculus, 4,000 years back to Eve, was one explanation of its origin. (The name Homunculus was applied by later authors.)

William Harvey, well known for his discovery in the late sixteenth century of the circulation of blood, considered embryonic development to be the formation of the complex organism from non-structured material by the process he named epigenesis. By the seventeenth century, natural philosophers theorized that embryogenesis, the seeming miracle of the animal body growing from an invisible seed, might happen by either of two rational systems: by preformation, the enlargement of a miniature structure in the seed, or by epigenesis, the building of the body like the building of a house, step by step, by the adding of material upon the previous form. Whether life forms and their evolution are directed by preformation or by epigenesis continued to be a major debate of biological science.

The preformationists could not explain how the miniature body got there, while the epigenesists could not think of an inheritable system that could guide the building of the body structure, resorting always to the divine, or some vague, transcendental force, often called vitalism. Of course, no one could imagine a molecular code, or the ultimate plausible if not demonstrable explanation of preformation: self-organization.

The school of mechanicists, who in the eighteenth-century tradition of Kantian logic, saw development proceeding in conformance with mathematical and physical laws, championed the epigenesis school of thought. The German school of *Entwicklungsmechanik* saw embryogenesis as the generation of form by the natural folding of expanding tissue (Roux 1894; Korschelt 1902; Rhumbler 1903). This doctrine dominated European experimental embryology in the second half of the nineteenth century (Butschli 1897; Bammes 1886; Gallardo 1896).

In the nineteen century, morphologists began to look at the origin of the structure of the body by tracing evolution from fish to man, successfully demonstrating how fin became limb, how the gill arches became the inner ear bones, but never achieving an understanding of the general process of morphological change. The sciences of embryology and anatomy were customarily presented as purely descriptive and human anatomy was depicted with accuracy from Vesalius to Gray, for the benefit of surgeons and artists.

Real experimental embryology began with the nineteenth-century school of anatomists in Germany and Switzerland who sought a mechanical explanation for the sequence of events from the egg to the embryo. Wilhelm His (1875; 1903), now remembered as the father of human embryology, was engaged in the latter half of the nineteenth century in a life-long project to solve the riddle of morphogenesis, the mechanism that organizes the cells to form the embryo. His sought to reduce biological phenomena to mathematics, physics and mechanics. Possessed with the belief in a purely logical explanation of life, he went so far as to enlist Eduard Hagenbach, a Swiss mathematician, to find a mathematical algorithm for living form.

It was His who invented the modern microtome, out of necessity to make thin sections required to reconstruct the shape of the embryo and thereby to reveal the secret of morphogenesis. He collected human embryos at every stage, made drawings of each microtome section on the slide and wax models of the enlarged slices, which were stacked, thereby reassembling the form. Thus, a series of enlargements of the embryo at each stage was produced in the form of twelve eighteen-inch wax models showing the growth pattern as a time-lapse series of sculptures. He held that the mechanical forces on the expanding tissue, which respond by simple laws of mechanics and hydrodynamics, directed form. He demonstrated that the proto-brain takes shape in mimicry of the behavior of a rubber tube subject to the same forces. This theory of mechanicism, or *Entwicklungsmechanik,* was a prevailing school of thought throughout the second half of the nineteenth century (Roux 1894). The continuation of this costly human embryogenesis project might eventually have revealed the underlying geometric machine.

It was not to be. Biology became dominated by the notorious Ernst Haeckel, professor of embryology at Jena, author of the biogenetic law that ontogeny recapitulates phylogeny (Haeckel 1866, 1896). The Wagnerian proto-Nazi, Haeckel, was Darwin's evangelist in Germany. He feared that *Entwicklungsmechanissmus* could refute Darwin and eventually his own theory. His archrival was Wilhelm His. It was His who exposed the famous fraud where Haeckel faked drawings of embryos to show commonality of descent. Using his tremendous prestige, Haeckel proceeded to destroy His and the whole school of thought. He ridiculed his work as *Gummischlauchwissenshaft*, "rubber tube science." His and his followers could not publish in the German journals nor get university appointments. The wax models may still be seen at the University of Basel.

The approach to biology called mechanism, or structuralism, denies the working of any of the divine, transcendental or extra-mechanical forces, which have been invoked in the failure of the simpler, direct, intuitively comprehensible effects common to other sciences. Aristotle isolated natural history from the rest of philosophy with the term entelechy, implying a life force. In frustration with the relentless resistance of nature to yield up a solution to the riddle, after a century of search for a mechanical solution to embryogenesis, many biologists as recently as the beginning of the last century embraced the principle of vitalism, put forth by the great embryologist Hans Driesch, who revived the term entelechy after two thousand years. In a 1907 lecture he declared:

No kind of causality based on the constellation of single physical and chemical acts can account for organic individual development.... Life, at least morphogenesis is not a specialized arrangement of inorganic events; biology therefore is not applied physics and chemistry. Life is something apart, and biology is an independent science.

At the end of a respected career including hundreds of noted landmark lectures and writings, Driesch renounced science and, finally, reason itself.

The great Scottish zoologist D'Arcy Thompson in his monumental, 1116-page work of 1917, *On Growth and Form*, the *summum* of the mechanistic theory of life, established the perfect concordance of the forms of species with simple, mathematically predictable configurations. The simple geometric figure upon which the body plan is based is evident in the single cell organisms and in the entire class of radial animals including the sponges, corals, jellyfish, starfish and crinoids. What need for a transcendental explanation of how a jellyfish is made when you may calculate it in advance by principles of topology, the geometry of surfaces, and mimic it as did Plateau by dripping ink in a vortex?

D'Arcy Thompson presents a convincing argument that evolution consists of the gradual elaboration of a fundamental, self-organized natural configuration, as suggested by Goethe. Stephen Jay Gould, in his introduction to the abridged edition of the work, characterizes D'Arcy Thompson's theory of biology as a hybrid of

Pythagoras and Newton, arguing, *Physical forces shape organisms directly, with internal and genetic forces responsible only for producing raw material in gradients and programmed timings, for construction under principles of physics.* D'Arcy Thompson's influence is still here today. "He gave simple engineers answer to the Darwinism that was the biology of his time. Thompson believed in an economy of hypotheses, an approach to problem solving known as Ockham's razor from the fourteenth-century philosopher, William of Ockham. Put simply, his approach demands that we set aside complicated explanations for things when a simpler one will do" (Ball 1999). While the book is hailed as one of the great works in the scientific literature, it is not taken seriously today as it obviously contradicts Darwin and the theory of genetic determination of form.

Cytoplasm vs. Nucleus, Organicism vs. Cell Theory, Mechanism vs. Vitalism: these debates prevailed in the first decades of the twentieth century. At issue was again the resolution of the age-old riddle of embryogenesis. The dispute now was whether the formation of the embryo, and thus the adult organism, was directed by the nucleus of the egg or by the cytoplasm in which it abides. The then Titans of biology, Edmund Beecher Wilson and Thomas Hunt Morgan, both professors at Columbia University, were the opposing protagonists in what turned into a vituperative combat, reminiscent of the great scientific contentions of the past. Morgan declared that Mendelian genetics held the secret of morphogenesis in a code. In a landmark experiment, Morgan removed part of the egg cytoplasm of the ctenophore, a marine invertebrate, prior to fertilization, and found that the embryo was accordingly defective, concluding that the egg cell prior to fertilization was already segmented into localized zones that led to the embryonic form. Morgan was for a time at least, the exponent of the cytoplasmic origin of embryogenesis. In another experiment, Morgan, using the ascidians, or sea squirt, a primitive marine chordate presumed to be the predecessor of the vertebrates, found that the removal of one of the eight blastomeres created a larva that lacked the respective portion, confirming the idea that the animal was a machine of separate, functioning parts, implying that the body is organized prior to fertilization (Morgan 1915, 1934). Contradictory results came to confuse the issue. Wilhelm Roux discovered that each of the first eight cells of the sea urchin blastocyst was able to create a complete animal, in refutation of the machine concept. Decades later, Morgan dramatically renounced the cytoplasm position, taking up the new science of genetics, representing the nucleus as the source of heredity, the death knell for the cytoplasmic position.

A standard text of the period, *Developmental Anatomy*, by embryologist Professor Leslie Arey, sums up the status of the conflict in the 1940 edition, thus:

The regional organization of the egg protoplasm represents a refined sort of preformation, but prelocalization is a more precise term. From these facts Conklin and Loeb argue that the cytoplasm is really the embryo in the rough; the nucleus through Mendelian heredity, adds only the finer details. Morgan and Lillie, refusing to admit the validity of this interpretation, regard the cytoplasm merely as a medium upon which the genes act.

The belated uncovering of the famous genetic rules of Gregor Mendel entered the equation, focusing attention on the genetic molecular material in the nucleus. Today's texts in developmental biology recount this important episode in the history of the establishment of the modern theory of genetics as a literal dogma of biology. Advances in the field of molecular biology came to reveal the seemingly unlimited potential of the gene to encode vast information, suggesting an all-embracing possibility for a mechanism for embryogenesis.

However, by 1930, the association of experimental embryology with Hitler's racism characterized deterministic biology as politically incorrect. But by the end of World War II, the triumph of genetics was complete. All biological phenomena were to be explained by the action of genes, and all characters and organs were to be seen as the result of advantageous adaptation to the environment. "Nature tried everything and this is what works," were the keywords.

Biologists are now beginning to question the supporting evidence for nature trying everything. Fossil records show animal forms in a straight line development of body plans that suddenly appeared after the Cambrian explosion some 543 million years ago and can find no evidence of abandoned body plans or organs of a different kind than in use now.

PLATE 59: *Darwin Contemplating Skull*

Chapter 25 Astrobiology and Universal Convergence

Alien biology would completely rewrite the earthly textbooks. In the abstract, however, there would still be a strong resemblance to life as it exists on Earth and again there would probably be little resistance to extending the definition of life to encompass these new creatures. (Ian Stewart 1998)

The infant science of astrobiology labors under the disadvantage of not as yet having accumulated any evidence to corroborate its own theory, that there is extraterrestrial life. It lives on hope. This is not due to any scientific principle, or evidence reasoning against it, as, for example, the case of parapsychology or homeopathy. To the contrary, the likelihood of extraterrestrial life is just a question of probability, and this increases with every new indication that the planets outnumber the stars, suggesting that there are many earthlike ones.

The current notion that the living body achieves its form as the result of environmental selection of random mutations, makes the prospect for the terrestrial type of life elsewhere seem hopeless, except in a duplication of the sequence of conditions that have prevailed on earth over a billion years. The difference between the terrestrial environment and almost any other would, by this reasoning, predicate forms vastly different than those we are familiar with, making it hard to imagine anything we could recognize as familiar life, no less communicate with. This is the argument of the detractors of astrobiology. Mathematicians and critics of SETI (Search for Extra Terrestrial Intelligence) have claimed that the likelihood of success is comparable to apes typing *Hamlet*. Such a factor keeps the probability near zero in the Drake equation, by which the number of civilizations in the Galaxy may be expressed.

There are an estimated over 400 million stars in the Milky Way galaxy alone. Probably at least one out of ten of these stars have planets equaling roughly 40 billion planets. If every star has ten planets, that's 400 billion planets. Even if we are conservative and estimate that only one planet in each stellar system might be ripe for life, that is still 40 billion habitable planets and that is just in the Milky Way, and there are an estimated one hundred billion galaxies like our Milky Way. The vast count of stars, virtually assures other earthlike environments. But even with the high likelihood of amino acids getting started making protoplasm by a Urey-Miller reaction, what is to assure that the random steps that follow will make a brain? A bottleneck in the Drake equation is morphogenesis. Yet it is now known that much of life

PLATE 60: *On the basis of the principle of universal morphological convergence, extraterrestrial life would resemble our own, as does the inorganic content of the universe. It could be argued that auto-evolution would inevitably produce a mollusk, arthropod or even a large-brained vertebrate, and it is therefore reasonable to attempt communication.*

follows simple physical laws. Denton and Marshal ask, "If forms as complex as the protein folds are intrinsic features of nature, forming only 500 to 1000 different varieties of proteins, might some of the higher architecture of life also be determined by physical law?" (Denton 1998, 2001)

The very principle of Darwinism argues against its ability to reproduce the complex forms of terrestrial life, given the slightest perturbation in the sequence of environmental events, which produced them over eons, driven by chance. Chief criticism of the quest for SETI has been that random morphogenesis leading to intelligent life forms is an incalculable pipe dream.

But fortunately for SETI there is no longer good reason to accept natural selection as a cause of phylogenesis, or macroevolution. Theoretical biologists have generally rejected the premise in the past several decades, attributing form to phenomena of self-organization, the principle behind inorganic crystals. The body is made by the self-organization of its tissues, as are the molecules of which it is composed. "Proteins follow simple physical laws and tissues can self-organize" (Kaufmann 1995). The eye, for instance, is known to appear over and over and over again in diverse forms. We may anticipate terrestrial type phyla on other worlds for the same reasons we anticipate terrestrial-type rocks. If non-heritable phyletic form were established, then extraterrestrial intelligence would be a near sure thing. On a sufficiently old earthlike planet the chance of finding a brain would be as great as that of finding snowflakes.

Meteorites and rocks from the moon and Mars have well established that inorganic matter is universal in form. The phenomena of chemistry and physics that govern are based on simple principles, observed to be universal. Such principles have been sought in vain for organic structure since Plato and Aristotle.

A fundamental question in the search for extraterrestrial life is what it would look like. The form of a living cell, or the form assumed by a multiplicity of them, may be ascribed to one of two opposing theories, which purport to account for the evolution of complex life forms. Natural selection means that the environment, in its selection of random mutations, shapes form. Orthogenesis or autogenesis means that form is determined from within, by fundamental properties of its constituents, like geological configurations and inorganic crystals. Protoplasm, a highly complex mixture of proteins is, of itself, capable of generating intricate internal structures and membranes of organized molecular arrays. Proteins form through simple biochemical and physical laws of nature. The cytoplasm of the egg is known to generate the rudiments of the embryonic form by self-organization. The question of whether the form of the embryo is directed by the genetic contents of the nucleus or by the structural potential of the cytoplasm was, and still is, an ongoing debate in theoretical biology.

Autogenetic theories of development claim that the construction of the embryo is directed by physical or mechanical causes of underlying mathematical order. Evidence for this premise is abundant in the obvious geometric organization of so much of the animal and vegetable worlds. The prevalence of the phenomenon called convergence, the appearance over and over again of the same forms in phyla with no common ancestry, is especially apparent in the similarity in body form of the higher bilateral animals. Even the finishing touches of evolution display convergence. From common ancestors the continents of America, Africa and Australia have each independently produced the same set of mammalian wildlife: that is, three sets of cats, dogs, bears, primates, differing only in infinitesimal refinements. Biodiversity consists of variations on one theme.

On the basis of the principle of universal morphological convergence, extraterrestrial life would resemble our own, as does the inorganic content of the universe (Plate 60). It could be argued that autoevolution would inevitably produce a mollusk, arthropod, or even a large-brained vertebrate, and it is therefore reasonable to attempt communication. The brain is, after all, nothing but a tube, cross-hatched with neurons. It was invented independently by each of the phyla and appears in the same form in all animals, complete and attached to the motor system. This fact has produced universal animal behavior. Much as the bilateral forms have one body plan, they also exhibit nearly identical behavior, based on territoriality, hunger and sex. The maneuvers of aggression and flight are ubiquitous in the phyla. This is why an octopus and a crab can engage in a fair fight, and why Mickey and Minnie Mouse can be so human. Intelligence depends only on the size of the brain. We may expect extraterrestrial carbon-based life to act like us.

Most biologists pursue their work taking for granted the central dogma of biology, that form is encoded in the genome and by the mutation of DNA can be randomly generated. While this is indeed the case in genotypic characters of the individual, forming the successful basis for applications in medicine and genetic engineering, the genome has offered little evidence of its direction of the phenotypic characters, the phyletic body form, common to all individuals of the phylum. The term blueprint of life is no longer applied to the genome. A considerable number, if not a majority of writers on evolutionary biology and embryology are skeptical of the role of natural selection except in microevolutionary effects.

While there is no proof of the effect of natural selection on macroevolution, science may yet look forward to the identification of an orthogenetically driven system of embryogenesis of the kind postulated over the centuries, and supported by leading evolutionary biologists of our era. This should offer encouragement to those who scan the universe for intelligent response, and give answer to critics who wrongly maintain the implausibility of extraterrestrial life in the terrestrial model, incorrectly presuming the random origin of form. If a self-organizational theory of embryology were established, then the odds that a life-friendly planet would produce ET are around the same as the production of snowflakes.

When I am judging a theory, I ask myself whether, if I were God, I would have arranged the world in such a way. (Einstein)

Bibliography

al-Adhami, M., Qar, J. & al Khodur, M. (2001) Embryonic Fissure and Photoreceptor Differentiation in the Eye of Adult Garra rufa Heckel 1843 (Cypinidae, Teleostei). *Folia Biol*, 49, 183–90

Alberch, P. & Kollar, E. (1988) Strategies of head development:Workshop Report. *Development*, 103 (Suppl.), 25–30

Alberch, P., Lewbart, G. & Gale, E. (1985) The Fate of Larval Chrondrocytes During the Metamorphosis of the Epibranchial in the Salamander Eurycea bislineata I. *Journal of Embryology and Experimenta Morphology*, 88, 71–83

Alberts, B., Johnson, Lewis, J. & Raff, M., Walter. (1997) *Essential Cell Biology*. United States: Garland Publisher

Albrecht-Buehler, G. (1992a) Speculation about the Function and Formation of Centrioles and Basal Bodies. In *The Centrosome*, ed. V. Kalnins, New York: Academic Press. 69–102.

Albrecht-Buehler, G. (1992b) A Rudimentary Form of Cellular 'Vision'. *Proc. Natl. Acad. Sci.USA*, 89, 8288–92

Albrecht-Buehler, G. (1994) The Cellular Infrared Detector Appears to be Contained in the Centro-some. *Cell Motility and the Cytoskeleton*, 27, 262–71

Albrecht-Buehler, G. (1995) Changes of Cell Behavior by Near-Infrared Signals. *Cell Motility and the Cytoskeleton*, 32, 299–304

Amirthalingam, K., Lorens, J., Saetre, B., Salaneck, E. & Fjose, A. (1995) Embryonic Expression and DNA-Binding Properties of Zebrafish Pax 6. *Biochemical and Biophysical Research Communications*, 215, 122–8

Angie, N. (1998) When Evolution Creates the Same Design Again and Again. *New York Times*, 15/December

Anson, R. (1951) *Atlas of Human Anatomy*. Philadelphia, PA: W.B. Saunders

Arber, A. (1950) *The Natural Philosophy of Plant Form* Cambridge University Press

Arey, L. B. (1940) *Developmental Anatomy*. PA: W.B. Saunders Company

Baggot, J. (1996) *Perfect Symmetry: The Accidental Discovery of Buckminsterfullerene*. Oxford, UK: Oxford, University Press

Ball, P. (1999) *The Self-Made Tapestry: Pattern Information in Nature*. Oxford, UK: Oxford University Press

Ballard, W. W. (1964) *Comparative Anatomy and Embryology*. New York, NY: Ronald Press

Bammes, G. (1986) *Die Gestalt Des Tieres*. Berlin: Ravensberg Press

Batterson, S. (2000) *Stephen Smale: The Mathematician Who Broke the Dimension Barrier*. Providence, RI: American Mathematical Society

Baumgart, T., Hess, S. T. & Webb, W. W. (2003) Imaging Coexisting Fluid Domains in Biomembrane Models Coupling Curvature and Line Tension. *Nature*, 425 (23/October), 821–4

Behe, M. (1996) *Darwin's Black Box*. New York, NY: TouchStone: Simon & Schuster

Bernays, E. (1986) Diet-Induced Head Allometry Among Foliage-Chewing Insects and Its Importance for Graminivores. *Europhys. Letters*, 13 (495–497)

Berndl, K., Kas, R., Lipowsky, R., Seifert, U., & Sackmann, E. (1990) Shape Transformations of Giant Vesicles: Extreme Sensitivity to Bilayer Asymmetry. *Science*, 231(659–64)

Bertalanffry, L. V. (1933) *Modern Theories of Development*. London, UK: Oxford University Press

Bohn, S., Andreotti, B., Douady, S., Munzinger, J. & Couder, Y. (2002) A Constitutive Property of the Local Organization of Leaf Venation Networks. *Phys. Rev. E*, 65 (June 28)

Bonner, J. (1993) *Life Cycles*. Princeton NJ: Princeton University Press

Briggs, J. (1992) *Fractiles: The Patterns of Chaos*. New York, NY: Simon & Schuster

Brookbank, J. (1978) Developmental biology: Embryos, plants, and regeneration, New York, NY: Harper and Row

Brunschwig, K., Wittmann, C. & Schnabel, R., Burglin TR. (1999) Anterior Organization of the Caenorhabditis Elegans Embryo by the Labial-Like Hox Gene ceh-13. *Development*, 126(7), April, 1537–46

Buchsbaum, R., Buchsbaum, M., Pearse, J. & Pearse, V. (1987) *Animals Without Backbones*. Chicago, IL: University of Chicago Press

Budd, G. (1999) Does Evolution in Body Patterning Genes Drive Morphological Change or Vice Versa? *BioEssays*, 21, 781–4

Buffon. (1749–1767) *Histoire Naturelle, Generale et Particuliere*. Paris, Fr.: Imprimerie Royale

Butschli, O. (1897) Ueber die Kunstliche Nachahnung der Karyokinetischen Figur. *Vers. Med Nat. Ver Heildelberg*, 5, 28–41

Canham, P. B. (1970) The Minimum Energy of Bending as a Possible Explanation of Biconcave Shape of Human Red Blood Cell. *J. Theor. Biol.*, 26, 61–81

Carroll, R. L. (1999) Homology Among Divergent Paleozoic Tetrapod Clades. In *Homology (Novartis Foundation Symposioum 222)*, ed. G. Bock & G. Cadew, Chichester, UK: Wiley. 47–64

Carroll, R. (1988) *Vertebrate Paleontology and Evolution*. New York, NY: W.H. Freeman and Co.

Carroll, S. B., Weatherbee, S. D. & Langeland, J. A. (1995) Homeotic Genes and the Regulation and Evolution of Insect Wing Number. *Nature*, 375, 58–61

Casares, F. & Mann, R. S. (1998) Control of Antenna Versus Leg Development in Drosophila. *Nature*, 392, 723–6

Chaline, J. (1990) *Paleontology of Vertebrates*. Berlin, Ger.: Springer-Verlag

Chen, T., Hiroko, T., Chaudhuri, A., Inose, F., Lord, D., Tanaka, S., Chant J. & Fujita, S. (1999) Pax6 Induces Ectropic Eyes in a Vertebrate. *Development*, 126, 4213–22

Cho, A. (2004) Life's patterns, No Need to Spell It Out. *Science*, 303 (6/February), 782–3

Chow, R. L., Altmann, C. R., Lang, R. A. & Hemmati-Brivanlou, A. (1999) Pax6 Induces Ectopic Eyes in a Vertebrate. *Development*, 126, 4213–22

Chun, C. (1903) *Zoologica dem Gesamtgebiete der Zoologie*. Stuttgart, Ger.: Verlag von Erwin Nagale

Cohen, J. (1995) Who do we blame for what we are. In *How Things Are*, ed. J. Brockman & K. Matson. London, UK: Weidenfield and Nicholson, 51–60.

Colbert, E. H. & Morales, M. (1991) *Evolution of the Vertebrates*. New York, NY: Wiley-Liss

Commoner, B. (2002) Unraveling the DNA Myth. *Harpers Magazine*, 304 (February). New York, NY. 39–47

Conklin, E. G. (1927) *General Morphology of Animals*. Princeton, NJ: Princeton University Press

Cook, T. A. (1979) *The Curves of Life*. New York, NY: Dover Publications, Inc.

Cooke, J. (1980) Compartments, "Switches," "Programs" and Vertebrate Development. *Nature*, 284, 216–7

Cowin, S. C. (2000) How Is a Tissue Built? *Journal of Biomechanical Engineering*, 122 (December), 553–69

Cuvier, G. (1834) *The Animal Kingdom in Four Volumes*. London, UK: Ludgate-Hill

Cuvier, G. (1835) Eloge de M. de Lamarck. *Memoires de l'Academie Royale Des Sciences de l' Institut de France*, 13, I–XXXI, XIII

Czerny, T. & Busslinger, M. (1995) DNA-binding and transactivation properties of Pax6: Three Amino Acids in the Paired Domain are Responsible for the Different Sequence Recognition of Pax6 and BSAP (Pax5). *Molecular and Cellular Biology*, 15, 2858–71

Darwin, C. (1859) *On the Origin of Species by Means of Natural Selection, or the Preservation of Favoured Races in the Struggle for Life*. London, UK: John Murray

Davidson, E. (2001) *Genomic Regulatory Systems: Development and Evolution*. San Diego, CA: Academic Press

Davidson, L. A., Oster, G. F., Keller, R. E. & Koehl, M. (1999) Measurements of Mechanical Properties of the Blastula Wall Reveal Which Hypothesizied Mechanisms of Primary Invagination are Physically Plausible in the Sea Urchin Strongylocentrotus purpuratus. *Developmental Biology*, 209, 221–38

Davidson, L., Koehl, M., Keller, R. & Oster, G. F. (1995) How do Sea Urchins Invaginate? Using Biomechanics to Distinguish Between Mechanisms of Primary Invagination. *Development*, 121, 2005-18

de Beer, G. R. (1930) *Embryology and Evolution*. Oxford, UK: Oxford Universal Press

de Beer, G. R. (1940) *Embryos and Ancestors*. Oxford, UK: Clarendon Press

de Beer, G. (1971) *Homology: An Unsolved Problem*. London, UK: Oxford, University Press

Dennet, D. (1995) *Darwin's Dangerous Ideas*. New York, NY: Simon & Schuster

Denton, M. (1986) *Evolution: A Theory in Crisis*. Bethesda, MD: Adler & Adler Publishers, Inc.

Denton, M. (1998) *Nature's Destiny*. New York, NY: Free Press

Denton, M. & Marshall, C. (2001) Laws of form revisited. *Nature*, 410 (3/22)

Desmond, A. (1992) *The Politics of Evolution*. Chicago, IL: University of Chicago Press

Dobzbansky, T. (1951) *Genetics and the Origin of Species*. New York, NY: Columbia University Press

Dover, G. (2000) *Dear Mr. Darwin*. London, UK: Phoenix Press

Dreisch, H. (1908) *The Science and Philosophy of the Organism I–II*. London, UK: Adams and Charles Black

Dubos, R. (1980) *Man Adapting*. New Haven, CT: Yale University Press

Duncan, D. M., Burgess, E. A. & Duncan, I. (1998) Control of Distal Antennal Identity and Tarsal Development in Drosophila by Spineless-Aristapedia, a Homolog of the Mammalian Dioxin Receptor. *Genes and Development*, 12, 1290–303

Edelman, G. (1988) *Topobiology*. New York, NY: Basic Books

Eldredge, N. (1999) *The Pattern of Evolution*. New York, NY: W. H. Freeman and Company

Eldridge, N. (1995) *Reinventing Darwin: The Great Debate at the High Table of Evolutionary Biology*. New York, NY: John Wiley & Sons

El Shafie, M., Sousa, M., Windt, J. & Kruger, T. (2000) *An Atlas of the Ultrastructure of Human Oocytes*. New York, NY: The Parthenon Publishing Group

Emlen, P. K. (2001) Costs and the Diversification of Exaggerated Animal Structures. *Science*, 291, 1534–6

Evans, J., Foster, J., McAvey, B. G., G.L., Kopf, G. & Schultz, R. (2000) Effects of Perturbation of Cell Polarity on Molecular Markers of Sperm-Egg Binding Sites on Mouse Eggs. *Biol. Reprod.*, 62, 76–84

Evans, E. A., (1980) Minimum Energy Analysis of Membrane Deformation Applied to Pipet Aspiration and Surface Adhesion of Red Blood Cells. Biophys. J. 30, 265–284

Farge, E. (2003) *Curr. Biol.*, 13, 1365–77

Farge, M (1992) Wavelet Transforms and Their Applications to Turbulence. *Annual Review of Fluid Mechanics* 24, 395–457

Fourcade, B., Mutz, M & Bensimon, D. (1992) Experimental and Theoretical Study of Toroidal Vesicles *Phys. Rev. Lett.* 68, 2551

Fusco, G., Brena, C. & Minelli, A. (2000) Cellular Processes in the Growth of Lithobiomorph Centipeds (Chilopoda Lithobiomorpha). A Cuticular View. *Zoologischer Anzeiger*, 239(91–102)

Gallardo, A. (1896) Essai d'Interpretation des Figures Caryocinetiques. *Anales del Museo de Buenos-Aires*

Garcia-Castro M. I., Vielmetter, E. & Bronner-Fraser, M. (2000) N-Cadherin, a Cell Adhesion Molecule Involved in Establishment of Embryonic Left-Right Asymmetry. *Science*, 288(5468), 12/May, 1047–51

Gardner, R. (1997) The Early Blastocyst is Bilaterally Symmetrical and its Axis of Symmetry is Aligned with the Animal-Vegetal Axis of the Zygote in the Mouse. *Development*, 124, 289–301

Gardner, R. (2001) Specification of Embryonic Axes Begins Before Cleavage in Normal Mouse Development of the Zygote in the Mouse. *Development*, 128, 839–47

Gardner, R. & Davies, T. J. (2003) The Basis and Significance of Pre-Patterning in Mammals. *Phil.Trans.R. Soc. Lond. B.*, 358 (2/July), 1331–9

Gardner, R., Surani, M. & Solter, D. (2003) Introduction to "Epigenesis Versus Preformation During Mammalian Development." *Phil. Trans. R. Soc. Lond. B.*, 358 (April), 1313–5

Gardner, R. L., Meridith, M. R. & Altman, D. G. (1992) Is the Anterior-Posterior Axis of the Fetus Specified Before Implantation in the Mouse? *Journal Experimental Zoology*, 264, 437–43

Garstang, W. (1929) The Origin and Evolution of larval forms. *Brit. Ass. Adv. Sci. Rep.*, 96th Mtg, 77–98

Garstang, W. (1922) The Theory of Recapitulation: A Critical Re-Statement of the Biognetic Law. *Journal of the Linnean Society (Zoology)*, 35, 81–101

Gehring, W. J. (2000) Reply to Meyer-Rochow. *Trends in Genetics*, 16, 245

Gehring, W. J. & Ikeo, K. (1999) Pax6 mastering eye morphogenesis and eye evolution. *Trends in Genetics*, 15, 371–7

Geoffroy Saint-Hilaire, E. (1822) Considerations generales sur la vertebre. *Memoires Du Museum d'Histoire Naturelle (Paris)*, 9, 89–119

Giard, A. (1907) J.B. Lamarck: Discours d'Ouverture. *Bulletin Scientifique de la France et de la Belgique*, 40, 12

Gilbert, S. (2000) *Developmental Biology*. Massachusetts: Sinauer Associates, Inc.

Gilbert, S. (2003) *Developmental Biology*. Sunderland, MA: Sinauer Associates

Gillmore, P. (1873) *Reptiles and Birds*. London, UK: Cassell, Petter & Galpin

Glardon, S., Holland, L. Z., Gehring, W. J. & Holland, N. D. (1998) Isolation and developmental expression of the Amphioxus Pax-6 Genegene (AmphilPax-6): Insights into Eye and Photoreceptor Evolution. *Development*, 125, 2701–10

Goethe, J. W. v. (1790) *Versuch die Metamorphose der Pflanzen zu Erklaren*. Gotha, Ger.: C. W. Ettinger

Goethe, J. W. v. (1817–1820) *Ur Naturwissenschaft Überhaupt, Besonders Zur Morphologie*. Stuttgart, Ger.: Cotta

Goethe, J. v. (1837) *Oeuvres d'Histoire Naturelle (Traduites par Ch.Fr. Martins)*. Paris, Fr.: Cherbuliez

Goodrich, E. (1930) *Studies on the Structure and Development of Vertebrates*. Chicago, IL: University of Chicago Press

Goodwin, B. (1994) *How the Leopard Changed Its Spots— The Evolution of Complexity*. New York, NY: Charles Scribner's Sons

Goodwin, B. (2000) The Life of Form. Emergent Patterns of Morphological Transformation. Comptes Rendus de l'Academie des Sciences, Paris. *Life Sciences*, 323(15–21)

Goodwin, B. & Kaufman, S. (1990) Spatial Harmonics and Pattern Specification in Early Drosophila Development. Part 1. Bifurcation Sequences and Gene Expression. *Journal of Theoretical Biology*, 144, 303–19

Gorder, P (2004) Vortex Drive. New Scientist Oct. 23.

Gould, S. (1977) *Ontogeny and Phylogeny*. Cambridge, MA: Belknap Press

Gould, S. J. (1977) *Ever Since Darwin*. New York, NY: W.W. Norton and Company

Gould, S. J. & Eldredge, N. (1977) Punctuated Equilibria: The Tempo and Mode of Evolution Reconsidered. *Paleobiology*, 3, 115–51

Green, P. B. (1999) Expression of Pattern in Plants: Combining Molecular and Calculus-Based Biophysical Paradigms. *American Journal of Botany*, 86(8), 1059–76

Greene, E. (1999) Phenotypic Variation in Larval Development and Evolution: Polymorphism, Polyphenism, and Developmental Reaction Norms. In *The Origin and Evolution of Larval Forms*, ed. B. Hall, K & M. H. Wake. San Diego, CA: Academic Press, 379–410

Grey, R. D., Wolf, D. P. & Hedrick, J. L. (1974) Formation and Structure of the Fertilization Envelope after Passage of the Eggs from the Coelom to Oviduct in Xenopus laevis. *J. Exp Zoology*, 201, 73–84

Guyenot, E. (1929) Le Mechanism de l'Evolutiion et l'Experience. In *L'Evolution en Biologie*, ed. Fondation "Pour la Science" Centre International de Synthese, pp. 35–58. Paris, Fr.: Alcan

Haeckel, E. (1866) *Generelle Morphogie der Organism*. Berlin, Ger.: Georg Reimer

Haeckel, E. (1896) *The Evolution of Man. A Popular Exposition of the Principle Points of Human Ontogeny and Phylogeny*. New York, NY: Appleton

Haeckel, E. (1904) *Art Forms in Nature*. Munich, Ger.: Prestel

Haldane, J. (1932) *The Causes of Evolution*. London, UK: Longman

Halder, G., Callaerts, P. & Gehring, W. J. (1995) Induction of Ectoic Eyes by Targeted Expression of the Eyeless Gene in Drosophila. *Science*, 267, 1788–92

Hall, B. K. (1999) *The Neural Crest in Development and Evolution*. New York, NY: Springer

Hamill O. P. & Martinac B. (2001) Molecular Basis of Mechanotransduction in Living Cells. *Physiol Rev*, 81(2), April, 685–740

Hamill, O. P. & McBride, D. W. J. (1992) Rapid Adaptation of Single Mechanosensitive Channels in Xenopus Oocytes. *Proceedings of the National Academy of Sciences*, 89(16), 15/August, 7462–55

Haraway, D. (1976) *Crystals, Fabrics and Fields*. New Haven, CT: Yale University Press

Hardy, W. B. (1900) On the Structure of Cell Protoplasm. *Journal of Physiology*, 24, 158–210

Harvey, E. (1956) *The American Arbacia and Other Sea Urchins*. Princeton, NJ: Princeton University Press

Heegaard, J., Carter, D. & Beaupre, G. (1999) Mechanically Modulated Cartilage Growth May Regulate Joint Surface Morphogenesis. *J Orthop Res*, 17, 509–17

Heinrichs, A. (2003) Gene Expression: Let's Do The Twist. *Nature Reviews Genetics*, 4(10), October, 760

Helfrich, W. (1973). Elastic Properties of Lipid Bilayers: Theory and Possible Experiments. *Z. Naturforsch.* C28:693–703

Herring, S. (1993) Formation of the vertebrate face: Epigenetic and Functional Influences. *American Zoologist*, 33, 473–83

Hilker, M. (2002) *Chemoecology of Insect Eggs and Egg Deposition*. Oxford, UK: Blackwell Publishing

His, W. (1875) *Unsere Korperform und Das Physiologische Problem Ihrer Entstehung*. Leipzig, Ger.: Verlag Von F.C.W. Vogel

His, W. (1903) *Lebenserinnerungen*. Leipsig, Ger.: Als Manuskript Gedruckt

Ho, M. & Saunders, P. (1979) Beyond Neo-Darwinism:An Epigenetic Approach to Evolution. *Journal of Theoretical Biology*, 78, 589

Holden, A. (1991) *Shapes, Space, and Symmetry*. New York, NY: Dover Publications Inc.

Holland, J. H. (1995) *Hidden Order: How Adaptation Builds Complexity*. Addison: Wesley Publishing Co.

Holland, P. (1999) The Future of Evolutionary Developmental Biology. *Nature*, 402 (December 2)

Home, E. (1975) Ultrastructural Studies of Development and Light-Dark Adaptation of the Eye of Coccinella Setempunctata L., with Particular Reference to Ciliary Structures. *Tissue Cell*, 7(4), 703–22

Hopwood, N. (1999) "Giving Body" to Embryos. *Isia*, 90 (September), 462–96

Houston, D. W. & Wylie, C. (2003) The Xenopus LIM-homeodomain protein Xlim5 Regulates the Differential Adhesion Properties of Early Ectoderm Cells. *Development*, 130(12), June, 2695–704

Huber, G. (1915) The Development of the Albino Rat, Mus Norvegicus Albinus. I. From the Pronuclear Stage to the Mesoderm Anlage: End of the First to the End of the Ninth Day. *Journal of Morphology*, 26, 247–358.

Hunt, G. W., Lord, G. J. & Peletier, M. A. (2003) Cylindrical Shell Buckling: A Characterization of Localization and Periodicity. *Discrete and Continuous Dynamical Systems Series B*, 3(4), November, 506–18

Hunt, G. (1983) Elastic Stability: In Structural Mechanics and Applied Mathematics. In *Collapse: The Buckling of Structures in Theory and Practice*, ed. J. Thompson & Hunt G.W., pp. 125–48. Cambridge, UK: Cambridge University Press

Huxley, J. (1924) Constraint of Differential Growth Ratios and Their Significance. *Nature*, 114, 895

Hyde, I. H. (1904) Differences in Electrical Potential in Developing Eggs. *The American Journal of Physiology*, XII(3), 241–75

Jahn, T. L. (1949) *Protozoa*. Dubuque, IA: William C. Brown Company

Jean, R. (1994a) *Phylotaxis: A Systemic Study in Plant Morphogenesis*. Cambridge, UK: University Press

Jean, R. (1994b) *Symmetry in Plants, Paleontology and Evolution*. Singapore: Princeton University Press

Jeong, J. Y., Mukhopadhyay, A. K., Dailidiene, D., Wang, Y., Velapatino, B., Gilman, R., Parkinson, A., Nair, G., Wong, B., Lam, S., Mistry, R., Segal, I., Yuan, Y., Gao, H., Alarcon, T., Brea, M., Ito, Y., Kersulyte, D., Lee, H., Gong, Y., Goodwin, A., Hoffman, P. & Berg, D. E. (2000) Sequential inactivation of rdxA (HP0954) and frxA (HP0642) Nitroreductase Genes Causes Moderate and High-Level Metronidazole Resistance in Helicobacter Pylori. *Journal of Bacteriology*, 182(18), September, 5082–90

Johnson, M. H. (2001) Mammalian Development: Axes in the egg. *Current Biology*, 11, R281–4

Johnson, M. H., Chrisholm, J. C., Fleming, T. P. & Houliston, E. (1986) A Role for cyto-plasmic determinants in the development of the mouse early embryo? In *Determinative Mechanisms in Early Development (Suppl. to Volume 97 of the Journal of Embryology and Experimental Morphology*, ed. C. Wylie & pp. 97–121. Cambridge: Company of Biologists

Johnson, P. (1991) *Darwin on Trial*. Washington, D.D.: Regentry Gateway

Jülicher, F. Seifert, U. & Lipowsky, R. (1993) Phase diagrams and shape transformations of toroidal vesicles *Classification Physics Abstracts* 82,70 1682

Kas, J. & E Sackmann, E. (1991) Shape Transitions and Shape Stability of Giant Phospholipid Vesicles in Pure Water Induced by Area-to-Volume Changes *Biophys. J.* 60 825–844

Kauffmann, S. (1993) *Origins of Order: Self-Organization and Selection in Evolution*. New York, NY: Oxford University Press

Kaufmann, S. (1995) *At Home in the Universe: The Search for the Laws of Self-Organization and Complexity*. New York, NY: Oxford University Press

Keller, E. F. (2000) *The Century of the Gene*. Cambridge, MA: Harvard University Press

Keller, E. F. (2002) *Making Sense of Life*. Cambridge, MA: Harvard University Press

Keller, R. (2002) Shaping the Vertebrate Body Plan by Polarized Embryonic Cell Movements. *Science*, 6/December

Keller, R., Davidson, L. & Shook, D. (2003) How We Are Shaped: The Biomechanics of Gastrulation. *Differentiation*, 71, 171–205

Keller, R., Shih, J. & Oster, G. (1991) Dynamical Models for Cell Rearrangement During Morphogenesis. In *Gastrulation: Movements, Patterns and Molecules*, ed. R. Keller, pp. 101–20. New York: Plenum Press

King, M.C. & Wilson, A.C. (1975) Evolution at Two Levels in Humans and Chimpanzees. *Science* Vol.188, pp.107-116

Kingsley, J. (1912) *Comparative Anatomy of Vertebrates*. Philadelphia: P. Blackstone's Son & Co.

Kmita-Cunisse, M., Loosli, F., Bierne, J. & Gehring, W. (1998) Homeobox Genes in the Ribbonworm Lineus sanguineus: Evolutionary Implications. *Proceedings of the National Academy of Sciences of the United States of America*, 95, 3030–5

Knabe, W. & Kuhn, H. (1997) Ciliogenesis in Photoreceptor Cells of the Tree Shrew Retina. *Anat Embryol (Berl)*, 196(2), August, 123–31

Korschelt, E. & Heider, K. (1902) *Lehrbuch der Vergleichenden Entwicklungsgeschichte*. Vienna, Austria: Verlag Von Gustav Fischer

Kumar, S. & Bentley, P. J. (2003) An Introductin to Computational Development. In *On Growth, Form and Computers*, ed. S. Kumar & P. Bentley, J. New York, NY: Elsevier Ltd. Academic Press

Lamarck, J. B. P. A. (1835) *Histoire Naturelle Des Animaux Sans Vertebres*. Paris, Fr.: J. B. Baillere

Larsen, E. (2003) Genes, Cell Behavior and the Evolution of Form. In *Origination of Organismal Form*, ed. G. G. Muller & S. A. Newman. Cambridge, MA: Bradford Books, 119–32

Lau, J., M., You, H. Z. & Yu, L. (2002) Lattice-Like Array Particles on Xenopus Oocyte Plasma Membrane. *Scanning Volume*, 24, 224–31

Laubichler, M. & Wagner, G. (2001) How Molecular is Molecular Developmental Biology? A Reply to Alex Rosenberg's Reductionism Redux: Computing the Embryo. *Biology & Philosophy*, 16, 53–68

Lawrence, P. A. (1992) *The Making of a Fly: The Genetics of Animal Design*. Oxford: Blackwell Scientific Publications

Lawrence, P. A. (2001) Science or Alchemy. *Nature Reviews Genetics*, 2, 139–42

Lecuit, T. & Wieschaus, E. (2000) Polarized Insertion of New Membrane from a Cytoplasmic Reservoir During Cleavage of the Drosophilia. *J Cell Biol*, 150, 849–60

Leeuwenhoek, September, 17. (1683) Containing Some Microscopical Observations, About Animals in the Scurf of the Teeth. In *Philosophical Transactions*. An Abstract of a Letter, pp. 14: 568–74

Lewin, R. (1999) *Complexity—Life at the Edge of Chaos*. Chicago, IL: University of Chicago Press

Lewis, E. (1992) Clusters of Master Control Genes Regulate the Development of Higher Organisms. *Journal of the American Medical Association*, 267, 1524–31

Lewis, E.B.; Nusslein-Volhard, C.; Weischaus, E. (1995) Nobel Prize of Physiology or Medicine 1955: The flies and the Keys of the Embryonic Development. *Medicina* 55(6):715-7.

Lewontin, R. (1998) *The Triple Helix*. London, UK: Harvard University Press

Lewontin, R. (2000) *It Ain't Necessarily So. The Dream of the Human Genome and Other Illusions*. New York, NY: New York Review of Books

Li, H. S., Yang, J. M., Jacobson, R. D., Pasko, D. & Sundin, O. (1994) Pax-6 is First Expressed in a Region of Ectoderm Anterior to the Early Neural Plate: Implications for Stepwise Determination of the Lens. *Developmental Biology*, 162, 181–94

Lillie, F. (1902) Maturation without Cleavage in the Egg of the *Annelid Chaetopterus* pergamentaceus, 24/March

Lillie, F. R. (1895) The Embryology of the Unionidae. *Journal of Morphology*, 10(1–100)

Lincecum, J. M., Fannon, A., Song, K., Wang, Y. & Sassoon, D. A. (1998) Msh Homeobox Genes Regulate Cadherin-Mediated Cell Adhesion and Cell-cell Sorting. *J Cell Biochem*, 70(1), 1/July, 22–8

Lipowsky, R. (1995) The morphology of lipid membranes. *Current Opinion in Structural Biology* 5, 531–540

Lipowsky, R. (1999) From Membranes to Membrane Machines. *Statistical Mechanics of Biocomplexity* ed. by D. Reguera, J.M. Rubi, and J.M.B. Vilar, "Lecture Notes in Physics", Berlin, Ger.: Springer-Verlag 527, 1–23.

Lipowsky, R. (1998) Vesicles and biomembranes. *Encyclopedia of Applied Physics* Vol. 23, VCH Publishers, Weinheim and New York. 199–222.

Louvet, S., Aghion, J., Santa-Maria, A., Mangeat, P. & Maro, B. (1996) Ezrin Becomes Restricted to Outer Cells Following Asymmetrical Division in the Preimplantation Mouse Embryo. *Developmental Biology*, 177(2), 1/August, 568–79

Lovtrup, S. (1983) Reduction and Emergence. *Rivista Di Biologia*, 76(3), 437–61

Lovtrup, S. (1984a) The Eternal Battle Against Empiricism. *Rivista Di Biologia*, 77(2), 183–209

MacDonald, J. & Anson, B. (1940) Variations in the Origin of Arteries Derived from the Aortic Arch, in American whites and Negroes. *American Journal of Physical Anthropology*, 27, 91–108

MacLise, J. (1848) *Comparative Osteology*. London, UK: Taylor and Walton

Mann, C. (1991) Lyn Margulis: Science's Unruly Earth Mother. *Science*, 252, 378–81

Marchant, J. (2000) Need a Hand? *New Scientist*, 12/August, 34–7

Margulis, L. & Sagen, D. (2000) *What is Life?* CA: University of California Press

Margulis, L. & Sagen, D. (2002) *Acquiring Genomes*. New York, NY: Basic Books

Mathew, W. (1915) *Tertiary Mammalia and Permian Vertebra*. New York, NY: American Museum of Natural History

Mayr, E. (1982) *The Growth of Biological Thought*. MA: The Belknap Press

McGinnis, W., Levine, M., Hafen, E., Kuroiwa, A. & Gehring, W. J. (1984) A Conserved DNA Sequence Found in Homeotic Genes of the Drosophila Antennapedia and Bithorax Complexes. *Nature*, 308, 428–33

McIver, T. (1988) *Anti-Evolution*. Baltimore: John Hopkins University Press

Medawar, P. & Medawar, J. (1977) *The Life Science: Current Ideas of Biology*. London, UK: Wildwood House

Meinhardt, H. (1995) *The Algorithmic Beauty of Sea Shells*. Germany: Springer-Verlag

Mendel, G. (1865) Experiments in Plant Hybridisation. *Proceedings of the Brünn Society for Natural History*

Milner, R. (1993) *The Encyclopedia of Evolution*. New York, NY: Henry Holt and Company

Milton, R. (1998) *Shattering the Myths of Darwinism*. VT: Park Street Press

Minelli, A. (2003) *The Development of Animal Form*. Cambridge, UK: Cambridge University Press

Miodownik, M. A. (2003) Using Mechanics to Map Genotype to Phenotype. In *On Growth, Form and Computers*, ed. S. Kuman & P. J. Bentley. New York, NY: Elsevier Academic Press, 203–19

Mohseni, K (2001) Statistical Equilibrium Theory for Axisymmetric Flows: Kelvin's Variational Principle and an Explanation for the Vortex Ring Pinch-off Process *Physics of Fluids* Vol 13(7) pp. 1924–1931.

Montagu, A. (1981) *Growing Young*. New York, NY: McGraw Hill Book Company

Moore, S. W., Keller, R. E. & Koehl, M. A. R. (1995) The dorsal involuting marginal zone stiffens anisotropically during its convergent extension in the gastrula of Xenopus laevis. *Development*, 121, 3131–40

Morgan, T.H. et al. (1915) *The Mechanism of Mendelian Heredity*. New York, NY: Holt Rinehart & Winston. Reprinted. Johnson Reprint Corporation with an Introduction by Garland E. Allen, 1978.

Morgan, T.H. (1934) Embryology and Genetics. New York: Columbia University Press

Morris, S. (2003) *Life's Solution: Inevitable Humans in a Lonely Universe*. Cambridge, UK: Cambridge University Press

Muller, D., Schoenenberger. (1997) Structural Changes in Native Membrane Proteins Monitored at Subnanometer Resolution with the Atomic Force Microscope: A Review. *J Struct Biol*, 119, 149–57

Muller, F. (1869) *Facts and Arguments for Darwin*. London, UK: John Murray, Albemarle Street

Muller, G. G. (2003) Homology: The Evolution of Morphological Organization. In *Origination of Organismal Form. Beyond the Gene in Developmental and Evolutionary Biology*, ed. G. B. Muller & S. A. Newman, pp. 51–70. Cambridge, Massachusetts: MIT Press

Muller, G. & Newman, S. (2003) Origination of Organismal Form: The Forgotten Cause in Evolutionary Theory. In *Origination of Organismal Form. Beyond the Gene in Developmental and Evolutionary Biology*, ed. G. B. Muller & S. A. Newman, pp. 3–10. Cambridge, MA: MIT Press

Muller, W. A. (1997) *Developmental Biology*. New York, NY: Springer-Verlag Inc.

Nagy, L. & Miodrag, G. (1999) Cell Lineages in Larval Development and Evolution of Holometabolous Insects. In *The Origin and Evolution of Larval Forms*, ed. B. K. Hall & M. H. Wake. San Diego, California: Academic Press, 275–300

Newman, S. (1993) Is Segmentation Generic? *BioEssays*, 110, 1–18

Newman, S. & Comper, W. (1990) "Genetic" Physical Mechanisms of Morphogenesis and Pattern Formation. *Development*, 119, 1–18

Nguyen, T. D., Chen & Johnson, D. P., J. R. (1998) Gene Structure Properties of TIGR, an Olfactomedin-Related Glycoprotein Clones from Glucocorticoid-Induces Trabecular Meshwork Cells. *J Biol Chem*, 273, 6341–50

Nieuwkoop, P. D. & Florschutz, P. A. (1950) Quelques caracteres speciaux de la gastrulation et de la neurulation de l'oeuf de Xenopus laevis Daud., et de quelques autres anoures, lere Partie—Etude descriptive. *Archives de Biologie*, 61, 113–50

Nijhout, H. (1990) Metaphors and the Role of Genes in Development. *Bioessays*, 12, 441–6

Nilsson, L. (1990) *A Child is Born*. New York, NY: Bantam Books

Nishii, I., Ogihara. (2003) A Kinesin, Inv A, Plays an Essential Role in Volvox Morphogenesis. *Cell*

Nyart, L. (1995) *Biology Takes Form: Animal Morphology and the German Universities*. Chicago, IL: Chicago University Press

Oparin, A. (1938) *Origin of Life*. New York, NY: Dover Publications

Oster & Alberch. (1982) Evolution and Bifurcation of Developmental Programs. *Evolution*, 36, 444–59

Oyama, S. (2000) *The Ontogeny of Information: Developmental Systems and Evolution*: Duke University Press

Pakkasmaa, S. & Piironen, J. (2001) Water Velocity Shapes Juvenile Salmonids. *Evolutionary Ecology*, 14, 721–30

Papineau, D. (1995) Natural Selections. *New York Times*, May 14. Sunday, Late Edition. New York, NY

Patten, B. M. (1951) *Early Embryology of the Chick*. New York, NY: McGraw-Hill Book Company

Pearson, H. (2002) News Feature. *Nature*, 418 (July), 14–5

Percival Smith, A., Weber, J., Gilfoyle, E. & Wilson, P. (1997) Genetic Characterization of the Role of the Two Hox Proteins, Proboscipedia and Sex Comb Reduced, in Determination of Adult Antennal, Tarsal, Maxillary Palp and Proboscis Identities in *Drosophila melangaster*. *Development*, 124, 5049–62

Peterson, K. J., Irvine, S. Q., Cameron, R. A. & Davidson, E. H. (2000) Quantitive Assessment of Hox Complex Expression in the Indirect Development of the Polychaete Annelid Chaetopterus sp. *Proceedings of the National Academy of Sciences of the United States of America*, 97, 4487–92

Phillips, J. B. (1975) *Development of Vertebrate Anatomy*. St Louis, IL: C.V. Mosby Company

Pineda, D., Gonzales, J., Callaerta, P., Ikeo, K., Gehring, W. & Salo, E. (2000) Searching for the protypic eye genetic network: Sine oculis is essential for eye regeneration in planarians. *Proceedings of the National Academy of Sciences of the United States of America*, 97, 4525–9

Piotrowska K. & Zernicka-Goetz, M. (2001) *Nature*, 409, 517–21

Pochan, D., Chen, Z., Cul, H., Hales, K., Qi, K., & Woolsey, K. (2004) Toroidal Triblock Copolymer Assemblies. Science, 306, 94–97.

Pollack, G. H. (2001) *Cells, Gels and the Engines of Life*. Seattle, WA: Ebner & Sons

Pratt, H. P. (1989) Marking time and making space: Chronology and topography in the early mouse embryo. *Int Rev Cytol*, 117, 99–130

Raff, R. A. (1996) *The Shape of Life, Genes, Development and the Evolution of Animal Form*. Chicago, IL: Chicago Universal Press

Rakyan, V. & Whitelaw E. (2003) Trans-generational Epigenetic Inheritance. *Current Biology*, 13(1), R6

Randinsky, L. (1987) *The Evolution of Vertebrate Anatomy*. St. Louis, MO: C.V. Mosby Co.

Resh, V. H. & Carde, R. T. (2003) *Encyclopedia of Insects*. New York, NY: Academic Press

Resnick, D. (1995) Developmental Constraints and Patterns: Some Pertinent Distinctions. *J. Theor. Biol.*, 173, 231–40.

Rhumbler, L. (1903) Mechanisch e Erklarung der Ahnlichkeit Zwischen Magnetischen. *Arch f Entwmech*, 26, 476–535

Ridley, M. (1999) *Genome: The Autobiography of a Species in 23 Chapters*. New York, NY: Harper Collins Publishing

Roberts, S. J., Leaf, D. S., Moore, H. P. & Gerhart, J. C. (1992) The Establishment of the Polarized Membrane Traffic in *Xenopus laevis* Embryos. *J Cell Biol*, 118, 1359–69

Roman, B., PMMH. (2003) Buckling Cascade in Thin Sheets and Leaves. In *Thin Plate and Fiber Mechanics in Biology*. Austin, TX, March 6

Rose, S. (1998) *Lifelines: Biology Beyond Determinism*. Oxford, UK: Oxford University Press

Rosen, D. E. (1984) Hierarchies and History. *In Evolutionary Theory: Paths Into the Future*, ed. J. W. Pollard. Chichester, UK: John Wiley & Sons Ltd. 77–97

Rothchild, N. (1992) *Bionomics*. New York, NY: Henry Hold and Company

Roux, W. (1894) The Problems, methods and scope of developmental mechanics. In *Biological Lectures of the Marine Biology Laboratory Woods Hole*. Boston, MA: Ginn. 149–90

Sackin, H. (1995) Stretch-Activated Ion Channels. *Kidney Int*, 48(4), 1134–47

Saffman, E. & Laske, P. (1999) Germline Develoment in Vertebrates and Invertebrates. *CMLS Cellular and Molecular Life Sciences*, 55, 1141–63

Salazar-Ciudad, I., Jernvali, J. & Newman, S., A. (2003) Mechanisms of Pattern Formation in Development and Evolution. *Development*, 130, 2027–37

Salser, S. & Kenyon, C. (1994) Patterning C. Elegans: Homeotic Cluster Genes, Cell Fates and Cell Migrations. *Trends in Genetics*, 10, 159–64

Sanson, B., White, P. & Vincent, J. (1996) *Nature*, 383, 627–30

Sathananthan, A. (1998) Paternal Centrosomal Dynamic in Early Human Development and Infertility. *J Assist Reprod Genet*, 3 (March 15), 129–39

Saunders, J. J. (1982) *Developmental Biology*. New York, NY: McMillan Publishing Co., Inc

Saunders, P. (1992) *Morphogenesis: Collected Works of A.M. Turing*. Netherlands: Elsevier Science Publishers B.V.

Schauerhammer, R. (2001) The Natural Science of Johann Wolfgang v. Goethe. *21st Century Science & Technology*, 14(1), 12–26

Schejter, E. D. & Wieschaus, E. (1993) Bottleneck Acts as a Regulator of the Microfilament Network Governing Cellularization of the Drosophia Embryo. *Cell*, 75, 373–85

Schlichting, C. D. & Pigliucci, M. (1998) *Phenotypic Evolution: A Reaction Norm*. Sunderland, MA.: Sinauer Associates

Schmitt, R. & Sumper, M. (2003) How to Turn Inside Out. *Nature*, 424 (July 31)

Schnackenberg, B. J., Khodjakov, A., Rieder, C. L. & Palazzo, R. E. (1998) The Disassembly and Reassembly of Functional Centrosomes in Vitro. *Cell Biology*, 95(16), 4/August, 9295–300

Schoenwolf, G. C. & Mathews, W. W. (2003) *Atlas of Descriptive Embryology*. Upper Saddle River, NJ: Pearson Education, Inc.

Scott, I. C. & Stainier, D. Y. (2003) Twisting the Body into Shape. *Nature*, 6957 (2/October), 461–3

Scott, L. A. (2000) Oocyte and Embryo Polarity. *Semin. Reprod. Med.*, 18(2), 171–83

Sharp, L. W. (1921) *An Introduction to Cytology*. New York, NY: McGraw Hill Book Company

Shubin, N. H. & Marshal, C. R. (2000) Fossils, Genes and the Origin of Novelty. In *Deep Time. Paleobiology's Perpective*, ed. D. Erwin & S. Wing. Paleobiology 26 (4Suppl). 324–40

Singers, J. (1983) Vibrations and Buckling of Imperfect Stiffened Shells—Recent Developments, ed. J. Thompson & G. Hunt. Cambridge, UK: Cambridge University Press. 443–79

Slack, J. (1983) *From Egg to Embryo: Determinative Events in Early Development*. Cambridge: Cambridge University Press

Slaney, A., Takatoh. (1998) Defect Textures in Liquid Crystals: The Optics of Thermotropic Crystals. In *The Optics of Thermotropic Liquid Crystals*, ed. S. Elston & R. Samples

Smale, S. (1959) A Classification of Spheres in Euclidean Spaces. *Annals of Mathematics*, 69 (February), 327–44

Smith, J. (1980) *Embryol. Exp. Morph.*, 55, 257–77

Snodgrass, R. (1993) *Principles of Insect Morphology*. Ithaca, New York: Cornell University Press

Sole, R. & Goodwin, B. (2000) *Signs of Life: How Complexity Pervades Biology*. New York, NY: Basic Books

Sonneborn, T. (1965) Cytoplasmic Inheritance of the Organization of the Cell Cortex in Paramecium Aurelia. *National Academy of Sciences*, 53(2), 278–82

Spemann, H. (1938) *Embryonic Development and Induction*. New Haven, CT: Yale University Press

Spemann, H. & Mangold, H. (1924) Über Induktion von Embryonalanlagen durch Implantation artfremder Organisatoren. *Archiv Fur Mikroskopische Anatomie und Entricklungsmechanik*, 100, 599

Spitz, F. & Duboule, D. (2001) The art of making a joint. *Science*, 291, 1713–4

Steele, C., Stanford University. (2003) Buckling Cascade in Thin Sheets and Leaves. In *The Role of Mechanical Forces in the Shoot Apical Meristem*. Austin, TX, March 6

Steele, E. J., Lindley, R. & Blanden, R. V. (1998) *Lamarck's Signature: How Retrogenes Are Changing Darwin's Natural Selection Paradigm*. Reading, MA: Perseus Books

Stewart, I. (1998) *Life's Other Secrets*. New York, NY: John Wiley & Sons, Inc.

Stewart, I. (2003) Broken Symmetries and Biological Patterns. In *On Growth, Form and Computers*, ed. S. Kuman & P. J. Bentley, pp. 181–202. New York, NY: Elsevier Academic Press

Stewart. I. (2001) *What Shape is a Snowflake*. New York, NY: W.H. Freeman and Company

Swartz, J. (1999) *Sudden Origins*. New York, NY: John Wiley & Sons, Inc.

Tam, P. P., Gad, J. M., Kinder, S. J., Tsang, T. E. & Behringer, R. R. (2001) Morphogenetic Tissue Movement and the Establishment of Body Plan During Development from Blastocyst to Gastrula in the Mouse., 6, June. 508–17

Tamm, E. R., Russel, P., Johnson, D. H. & Piatigorsky, J. (1996) Human and Monkey Trabecular Meshwork Accumulates Alpha Beta Crystallin in Response to Heat Shock and Oxidative Stress. *Invest Ophthalmol Vis Sci*, 37, 2402–13

Tamm, E. R., Russell, P., Epstein, D. L., Johnson, D. H. & Piatigorsky, J. (1999) Modulation of Myocilin/TIGR Expression in Human Trabecular Meshwork. *Invest Opthalmol Vis Sci*, 40, 2577–82

Tartar, V. (1961) *The Biology of Stentor*. Oxford, UK: Pergamon Press

Thompson, D. W. (1917) *On Growth and Form*. Cambridge, UK: Cambridge University Press

Thompson, J. (1982) *Instabilities and Catastrophes in Science and Engineering*. New York, NY: John Wiley & Sons

Thompson, J. (2001) *Chemistry and Life Science*. Cambridge, UK: Cambridge University Press

Tickle, C. (2003) *Patterning in Vertebrate Development*. New York, NY: Oxford University Press

Tomarev, S. I., Callaerts, P., Kos, L., Zinovieva, R., Halder, G., Gehring, W. & Piatigorsky, J. (1997) Squid Pax-6 and Eye Development. *94, Proceedings of the National Academy of Sciences of the United States of America*, 2421–6

Trougakos, I. P. & Margaritis, L. H. (2002) Novel Morphological and Pysiological Aspects of Insect Eggs. In *Chemoecology of Insect Eggs and Egg Deposition*, ed. M. Hilker & T. Meiners. Malden, MA: Blackwell Publishing. 3–36

Tudge, C. (2000) *The Variety of Life*. Oxford, UK: Oxford University Press

Turing, A. M. (1952a) The Chemical Basis of Morphogenesis. *Phil. Trans. R. Soc. Lond. B*, 237, 37–72

Turing, A. M. (1952b) Butterfly Mimicry: The Genetical Evolution of an Adaptation. *Evol. Biol.*, 10, 163–226

Uetake, Y., Kato, K., Washitani-Nemoto, S. & Nemoto, S. S. (2002) Nonequivalence of Maternal Centrosomes/Centrioles in Starfish Oocytes: Selective Casting-Off of Reproductive Centrioled into Polar Bodies. *Dev. Biol.*, 247(1), July1, 149–64

Van Auken, K., Weaver, D. C., Edgar, L. G. & Wood, W. B. (2000) *Caenorhabditis elegans* Embryonic Axial Patterning Requires Two Recently Discovered Posterior-Rroup Hox Genes. *Proceedings of the National Academy of Sciences PNAS*, 97(9), 25/April, 4499–503

Van der Hoef, M., Dictus, W., Hage, W. & Bluemink, J. (1984) The Untrastructural Organization of Gap Junctions Between Follicle Cells and the Oocyte in Xenopus Laevis. *Eur. J. Cell Biol.*, 33, 242–7

Vitruvius. (25 BCE) *Ten Books of Architecture*

Vogel, K. G. & Koob, T. J. (1989) Structural Specialization in Tendons Under Compression. *Int Rev Cytol*, 115, 267–93

Vogt, K. C. V. (1851) *Zoologische Briefe*. Frankfurt, Germany: Literarische Anstalt

Vogt, W. (1929) Gestaltungsanalyse am Amphibienkeim mit ortlicher Vitalfarbung. *Wilthelm Roux Archiv Für Entwicklungs-mechanik*, 120, 384–706

Waddington, C.H. (1956) *Principles of Embryology*, op. cit., p. 412

Wagner, P. A., Chiu, C. H. & Laubichler, M. (2000) Developmental Evolution as a Mechanistic Science: The Inference from Developmental Mechanisms to Evolutionary Processes. *American Zoologist*, 40, 819–31

Wainwright, S. A. (1988) *Axis and Circumference: The Cylindrical Shape of Plants and Animals*. Cambridge, UK: Harvard University Press

Wallace, A. (1997) *The Origin of Animal Body Plans*. Cambridge, UK: Cambridge University Press

Wang, N., Butler, J. P. & Ingber, D. E. (1993) Mechanotransduction Across the Cell Surface and Through the Cytoplasm. *Science*, 21(260), 21/May, 1124–7

Warner, D. (1964) A Suggested Conformation for the Protein Sub-Unit of Tobacco Mosaic Virus. *J. Theor. Bio.*, 6, 118–36

Watson, J.L. & O'Farrell, A. F. (1991) Odonata (dragonflies and damselflies). In *The Insects of Australia Vol. 1*, ed. E. Neumann, P. Carne, E. Lawrence, J. Spradbery, R. Taylor, M. Whitten & M. Littlejohn. Melbourne, Australia: Melbourne University Press. 294–310

Watson, J. & Crick, F. (1953) A Structure for Deoxyribonucleic Acid. *Nature*, 171 (2/April), 737

Watson, P. A. (1991) Function follows form: Generation of intracellular signals by cell deformation. *Fed Am Soc Exp Biol*, 5, 2013–9

Wesson, R. (1994) *Beyond Natural Selection*. Cambridge: MIT Press

West-Eberhard, M. J. (2003) *Developmental Plasticity and Evolution*. Oxford, UK: Oxford University Press

Wickham, M. & Worthen, D. (1976) Centrioles and Cilia in the Mesothelial Cells of the Pericanalicular Region, June 15. 499–502

Wierzbicki, T. (1983) On the Formation and Growth of Folding Modes. In *Collapse: The Buckling of Structures in Theory and Practice*, ed. J. Thompson & G. Hunt. Cambridge, UK: Cambridge University Press. 19–30

Williams, G. C. (1966) *Adaptation and Natural Selection: A Critique of Some Current Evolutionary Thought*. Princeton, NJ: Princeton University Press

Williams, R. (1979) *The Geometrical Foundation of Natural Structure*. New York, NY: Dover

Williamson, D. I. (2003) *The Origins of Larvae*. Boston, MA: Kluwer Academic Publishers

Willmer, P. (2002) Convergence and Homoplasy in the Evolution of Organismal Form. In *Origination of Organismal Form. Beyond the Gene in Developmental and Evolutionary Biology*, ed. G. B. Muller & S. A. Newman, pp. 33–50. Cambridge, MA: MIT Press

Wilson, E. T., Cretekos, C. & Helde, K. A. (1995) Cell Mixing During Early Epiboly in Zebra Fish Embryo. *Developmental Genetics*, 17, 6–15

Wolpert, L. (1969) Positional Information and the Spatial Pattern of Cellular Differentiation. *Journal of Theoretical Biology*, 25, 1–47

Wolpert, L. (1991) *The Triumph of the Embryo*. Oxford, UK: Oxford University Press

Wolpert, L. (2003) Relationships Between Development and Evolution. In *On Growth, Form and Computers*, ed. S. Kuman & P. J. Bentley. New York, NY: Elsevier Academic Press. 47-63

Wolpert, L., Beddington, R., Jessell, T., Lawrence, P., Megorowitz, E. & Smith, J. (2002) *Principles of Development*. New York, NY: Oxford Press

Worden, G. (1984) *Mütter Museum of the College of Physicians of Philadelphia*. New York, NY: Blast Books

Zernicka-Goetz, M. (2002) Patterning of the embryo: The First Spatial Decisions in the Life of a Mouse. *Development*, 129(4), 815

Zimmer, C. (2001) *Evolution: The Triumph of an Idea*. New York, NY: Harper Collins Publishing

Acknowledgements

Editorial:
K. Hall, Ph.D. and S. Soter, Ph.D.

Anatomists/Artists/Sculptors:
Peter Sheesley
Stephen Halker
Sammy Shin
Ryan Toth

Photography:
Melissa Ludwig

Back Cover:
Model of early amphibian, Acanthostega
Eliot Goldfinger, Sculptor
coll. American Museum of Natural History

Design and production:
Scott Santoro, Worksight

Index